Fire Protection

Robert C. Till • J. Walter Coon

Fire Protection

Detection, Notification, and Suppression

Second Edition

 Springer

Robert C. Till
John Jay College of Criminal Justice
New York, NY, USA

J. Walter Coon
Overland Park
Kansas, USA

ISBN 978-3-030-08113-3   ISBN 978-3-319-90844-1   (eBook)
https://doi.org/10.1007/978-3-319-90844-1

For My Parents

Preface

> While fire protection engineering should be left to the experienced professional, those outside this discipline should have sufficient knowledge not only to understand the fire protection engineer's input, but to help make the installation productive and cost effective.
>
> From the Foreword of the First Edition

This book is designed to educate third and fourth year fire science students, graduate students in protection management, mechanical engineers, architects, estimators, fire service personnel, and designers in the "nuts and bolts" of fire protection system selection, design, and equipment.

This cohort requires knowledge of the pros and the cons of what is being proposed and how systems should be compared to one another. It also gives non-fire engineering practitioners a sense of proportion when they are put in a position to select a consultant, and to give a sense of what the consultant may be doing and how a system is being matched to the hazard.

To match fire systems to hazards requires that one understands how detection and suppression systems work, what the hazards of the occupancy or process being protected are and understand what the ultimate goals are. Goals may include life protection, property protection, historic occupancy protection, etc. These areas of exploration will help one to decide what system to choose for the occupancy, and the codes and standards that need to be applied. Automatic systems are simply systems that do not require human intervention. An automatic sprinkler system will apply water to a fire without human intervention provided that there are no elements within the system that will cause it to fail.

The title of the first edition of this book, published in 1991 was *Fire Protection – Design Criteria, Options, Selection*. The book was and still is designed to provide a basic foundation for fire suppression, detection, and alarm systems.

Over the last 27 years some technology has changed dramatically. For example, the rise of the internet has changed the way devices speak to each other in a fire alarm system. Global warming and climate change have changed the way that gases are employed to suppress fire. The jurisdictional individuals, agencies, and standards involved in the field have also changed radically, with three major building codes integrated into the International Building Code first published in 1997. Finally, knowledge of possible fire department intervention and the time that it might require are being integrated.

All systems mentioned are governed by NFPA codes and standards which are available to the general public at no cost at: https://www.nfpa.org/Codes-and-Standards/All-Codes-and-Standards/Free-access.

FM Global data sheets are available at: http://www.fmglobal.com/research-and-resources/fm-global-data-sheets. Finally, while direct access to the current IBC is still an expense, access to the New York City Building Code, based on the 2014 version of the International Building Code (but with many modifications), is available at no cost at: https://www1.nyc.gov/site/buildings/codes/2014-construction-codes.page#bldgs. It is an excellent resource for educational purposes.

There are three simple steps to putting out a fire – fire detection, location, suppression.

1. Fire and Its Detection
2. Notification of Occupants and Fire Service
   - Occupants need to evacuate
   - Fire service needs to find the fire
3. Manual Fire Suppression
4. Automatic Suppression Systems

The book is divided such that each of these is covered by their own parts. The reader first must become aware of what fire is and how it grows. This stage is covered in Unwanted Fire and Fire Growth. Building inhabitants and the fire service both need to be notified so they can act accordingly. This stage is covered in Notification.

A major advance from the bucket brigade is the implementation of pumps and pipes to get water to the fire. This is included in fire pumps and water supplies. Finally, the suppression stage in general (including the fire service) is covered in Fire Suppression Systems. Fire suppression systems usually perform the act of both notification and suppression simultaneously, except in the case of manual suppression, where the fire department must act to deliver water to the fire.

A Word on Units

Time has been called the yardstick of fire control. The larger the fire is, generally the more heat and smoke are delivered in a given period of time. It is ironic that for most cases, heat release rates are given in metric units, so it might be more appropriate to say the time is the "meter-stick" of fire control.

Thus, the first chapter and in other chapters presenting information on heat release, equations are in English and metric units, while the later chapters will present them in imperial units, as is the custom for hardware here in the USA.

New York, NY, USA   Robert C. Till

Acknowledgments

First and foremost, I would like to thank my coauthor J. Walter Coon, PE, who I never had the chance to meet but certainly had the opportunity to learn a great deal from. I truly appreciate the permission of his children to continue the work.

I would like to thank Robert Fitzgerald, my good friend and PhD advisor at WPI. I never get to complain to Fitzy about how difficult writing can be, since he has always been writing something for as long as I have known him. Without his inspiration this effort would not have been completed.

I would also like to thank Doug Nadeau, Chris Marrion, Fred Hart, and Tom Derienzo for contributing photographs and for their general encouragement. I would also like to thank Tom and Robbie Snelham for all the coffee.

Finally, I would like to thank Paul Drougas and Caroline Flanagan at Springer for all their help in putting this together. My apologies to anyone I have missed.

Contents

1 Unwanted Fire and Fire Growth 1
  What is Fire? 1
    Fire Growth 1
  Fire in the Room of Origin 1
    Stages of Fire Growth 1
    Full Room Involvement 3
  Notification of Occupants 4
  Notification of the Fire Department 4
  Influence of Fire on Humans 4
  Modeling Fire: Empirically and Numerically 5
    Fire Triangle/Tetrahedron and Suppression 5
    Classes of Fire: What Is Burning? 5
    Calculation of Flashover 5
  Fire and Time: Fire Growth for a Single Fuel Package 6
    Diagnostic Room Fires 6
    Office Building Design Fires: Examples 7
    Open Office Plans and Spreadover 8
  Suppression of the Fire 10
  Summary 10
  References 10

2 Automatic Sprinkler Heads 11
  Early Automatic Suppression Systems 11
  History of the Sprinkler: 1864 12
    Fusible Element Sprinkler Head 12
    Parmelee Sprinkler Head 12
    Skepticism 12
    The Influence of Frederick Grinnell 13
  The Modern Spray Sprinkler Head 14
    Development of the Spray Sprinkler 14
    Sprinkler Heads 14
    Component: The Deflector 14
  Components: The Operating Element(s) 15
    The Solder Link 16
    Frangible Bulb Sprinkler Head 16
    Temperature Rating 16
    K-Factor 16

Development of the Large Orifice Head   17
Small Orifice Head   18
Dry Pendant Sprinklers   18
Sidewall Sprinklers   19
Decorative Sprinklers   20
Old Style Versus New Style Sprinkler Head   21
Special Heads   21
Large Drop Sprinkler   22
Early Suppression Fast Response (ESFR) Sprinkler   22
On-Off Sprinkler   22
Corrosion-Resistant Sprinklers   22
Rack Sprinklers   23
CMDA (Standard Spray)   23
CMSA (Large Drop)   24
Sprinkler Accessories   24
Other Common Characteristics of Spray Sprinkler   24
Summary   25
References   25

3   **Other Detection and Alarm Devices**   27
Detection and Alarm Systems   27
Alarm Initiating Devices   27
Manual Pull Stations   27
Thermal Detection Systems   27
Pilot Head Systems: Sprinklers Themselves   29
Continuous Line Detector   29
Protectowire Type Detector   29
Rate of Rise Detector   29
Ionization Versus Photoelectric Spot Detection   31
Ionization Detector Operation   31
Photoelectric Detector Operation   31
Linear Beam Smoke Detector   32
Rate Compensation Detector   33
Duct Detectors   34
Flame Detectors   34
Air Aspirating Smoke Detection Systems   37

4   **Notification**   39
History of Notification   39
The Telegraph   39
Fire Alarm   39
The Influence of John Gamewell   40
Telephone Systems   40
911 System   40
Internet and Addressable Systems   41
Notification System   41
Fire Alarm Systems   41
Fire Alarm Control Unit   41
Conventional FACU   41

Addressable FACU   41
Analog Addressable System   42
Supervision   42
Other Issues.   42
Types of Fire Alarm Systems   43
Local Alarm System   43
Auxiliary Alarm System   43
Remote Alarm.   43
Proprietary Alarm System.   44
Central Station   44
References.   44

5   Fire Pumps and Water Supplies   45
Early Methods of Suppression of Fires and Conflagrations   45
Fire Pumps   45
General Criteria.   46
Horizontal Fire Pumps   47
Suction and Discharge   47
Pump Suction   47
Suction and Discharge Piping.   48
Circulation Relief Valve   49
Main Relief Valve   49
Pump Test   50
Pump Test Design   50
Valve Supervision   52
Automatic Air Release   52
Diesel Fuel System   52
Pump Rotation   52
Engine Exhaust   52
Ventilation.   53
Diesel Engine Cooling   53
Reliable Power Supply to Electric Drive Fire Pumps.   53
Diesel Engine Exhaust   53
Diesel Engine Cooling System   53
Booster Pumps   54
Fire Pumps in a Bypass.   54
Pump House   54
Coupling Guards.   54
Jockey Pump Controller   54
Controller   55
Pressure Considerations   55
Controller Cabinet.   55
Controller Alarm and Signal Devices   55
Weekly Diesel Drive Pump Test   56
Fire Pump Controller Alarm Signal Operations   57
Diesel Engine Batteries.   57
Vertical Turbine Fire Pumps   57
Flushing and Testing.   57
Fire Pump Field Acceptance Test   57

Miscellaneous   59
Water Supplies   59
   Tank Selection.   59
   Concrete Reservoir   60
   Embankment Fabric Reservoir   61
   Tank Heating.   61
   Tank Water Level   61
Summary.   61

6 Underground Fire Mains   63
Underground Fire Mains.   63
Underground Pipe Selection   63
   Cast Iron Pipe   63
   Ductile Iron Pipe.   64
   Specifics for Cast Iron and Ductile Pipe.   64
   Friction Loss: The "C" Factor.   64
   Other Types of Underground Piping.   65
   Polyvinyl Chloride Plastic Pipe (PVC).   65
   Fiberglass-Reinforced Plastic Pipe   65
   Asbestos Cement Pipe.   65
   Steel Pipe   66
   Underground Pipe Bury   66
   Rodding and Thrust Blocks   66
   Other Considerations   67
   Hydrants   67
   Flushing and Testing.   67

7 Equipment and Devices   69
Equipment and Devices   69
   Control Valves.   69
   The Outside Screw and Yoke   69
   The Post Indicator Valve (PIV).   69
   Important Considerations When Using the PIV   71
   Butterfly Valve   71
   Pit OS&Y and Post Indicator   72
   Check Valves.   72
   Detector Check Valves and Full Flow Fire Meters   72
   Key Valve   73
   Full Flow Fire Meters   73
   Backflow Preventers   73

8 Firefighter Intervention: Manual Fire Suppression   75
The Fire Service: Introduction   75
   Established Burning to Arrival   75
   Arrival and Size Up.   76
Manual Suppression Coordinated Operations   78
   Locating the Fire.   78
Establish a Continuous Water Supply.   78
Standpipes: Minimizing Set Up Time.   78

Class I Standpipe. 79
Water Supply. 79
Pressure Requirements 79
Class II Standpipe 79
Water Supply. 80
Pressure Requirements 80
1–1/2″ Hose Stations. 81
Class III Standpipe 81
Standpipe Water Supply 82
Types of Standpipes 82
Hose Station Distribution 83
Roof Hose Stations. 85
Standpipe Zoning 85
Combined Standpipe and Sprinkler Riser. 86
Water Supply. 86
Set Up Time: Firefighting Tasks. 86
Set Up Time: Firefighter Factors. 87
Set Up Time: Location of Fire Within a Structure 87
Interior Attack. 87
Size of a Fire that Can Be Extinguished. 87
Average First Alarm Crews. 88
References. 88

9 Sprinkler Systems and Their Types 89
Water Agent Suppression Systems 89
Wet Pipe Sprinkler System 89
Advantages and Disadvantages. 91
Wet Pipe System Alarm 91
Water Damage. 91
Initiating the Alarm. 92
Water Flow Indicator. 92
Alarm Valve 94
Water Motor Alarm. 95
Pressure Switch. 95
Retard Chamber 95
Alarm Valve Trim 96
Fire Department Pumper Connection 97
Main Control Valve. 98
Inspector's Test 99
Discharge 100
Location 100
Wet Pipe System Drainage 100
Dry Pipe Sprinkler System 101
Advantages and Disadvantages of Dry Pipe Systems. 102
Differential Dry Pipe Valves. 103
Low Differential Dry Pipe Valve. 103
Dry System Air Supply. 104
Optional Devices. 105

Pressurized Nitrogen Supply. 106
Dry Valve Intermediate Chamber 106
Dry System Fire Department Pumper Connection 106
Quick Opening Devices 106
Accelerators 106
Exhausters. 107
Dry Pipe System Drainage 108
Pipe Support 108
Refrigerated Areas 108
Special Conditions 108
Maintenance Considerations. 108
Drainage Capacities 108
Drum-Drip Drain 109
Low Points 109
Dry System Design Features. 109
Calculating the Area of Application 109
Location of Dry Valves. 110
Water Columning 110
Deluge Systems. 111
Pre-action Systems 113
The Pre-Action Valve 114
Detection System 114
Protection of Equipment. 116
Supervising the Piping System with Air. 116
Use of Preaction Systems 116
Pre-action System Design Features 117
Types of Pre-action Systems. 118
Basic Pre-action System: Single Interlock 118
Double Interlock 118
Non-interlock 119
Water Spray System 119
Extinguishment. 120
Controlled Burning 120
Exposure Protection 121
Fire Protection. 121
Electrical Conductivity. 121
Water Spray Versus Sprinkler Systems. 121
Design Considerations 122
Nozzle Selection 122
Configuration of Equipment 123
Water Supply. 123
Water Mist Systems 124
Extinguishing Method. 124
Electrical Conductivity. 124
Design Considerations 124
Water Supply. 124
References. 124

10 Hydraulic Calculations of Sprinkler Systems 125
    Hydraulic Calculations 125
        Design Criteria 125
        Flow Data 126
        Flow Test... 127
        Approximate Calculated Demand....................... 129
    Computer Versus Hand Calculations 129
    Necessity of Hydraulic Calculations....................... 130
    Summary.. 130

11 Foam Systems 147
    History: Foam Systems................................... 147
    Use and Calculation of Foam 148
    Proportioning 149
        Selecting Proportioning Systems and Devices 149
    Balanced Pressure Proportioning 150
    Reserve Systems 150
    Direct Orifice Proportioning.............................. 151
    Detection Systems...................................... 151
    Jockey Pump.. 152
    Diaphragm Pressure Proportioning......................... 152
    Water Supply.. 152
    High Expansion Foam.................................... 153
        High Expansion Foam Generators 153
        High Expansion Foam Fire Suppression.................. 154
        Blessing and Curse 154
        High Expansion Foam Uses 154
        Sizing 155
        Advantages 155
        Activation 155
        Air Supply.. 155
    Types of Foam – Properties Performance.................... 155
        Protein Foam.. 156
        Fluoroprotein Foam (FP) 156
        Aqueous Film Forming Foam (AFFF) 156
        Alcohol-Resistant Foam Concentrate (AR-AFFF)........... 156
        Medium Expansion Foam.............................. 156
        High Expansion Foam................................. 156

12 Dry-Agent Automatic Suppression Systems 159
    Dry Agent Suppression Systems........................... 159
    Carbon Dioxide Systems................................. 159
        Local Application Systems 160
        Other Considerations 160
        Temperatures.. 160
        Grounding.. 160
    Total Flooding Systems.................................. 161
        Hazards of Total Flooding Carbon Dioxide 161
        Discharge Time Delay................................. 161

Re-Ignition   161
Choosing the Time Delay   162
Extended Discharge   162
Other Considerations   163
Carbon Dioxide and System Criteria   163
High Pressure Systems   163
Maintenance Considerations for High Pressure   164
Reserve Cylinders.   165
How the System Works.   165
Design of the Piping System.   166
Terminal Pressures   166
Discharge Nozzles.   166
Thumbnail Calculation   167
Low Pressure Carbon Dioxide Systems   167
The Storage Tank and Refrigeration.   167
Pipes and Fittings   168
How the System Works.   168
Dual Purpose Carbon Dioxide Units.   168
Inerting an Area with Carbon Dioxide   169
Carbon Dioxide Hose Stations   169
The Detection System and Air Leakage   170
Carbon Dioxide System Design   170
Local Application System Design.   170
Rate-By-Area Method.   170
Rate-By-Volume Method   172
Special Considerations   173
Total Flooding System Design   173
General Design Notes.   175
Clean Agents-Replacing Halon.   175
Types of Clean Agents   176
Greenhouse Gases and the Ozone Layer   176
Global Warming Potential.   176
Naming Conventions.   177
Inert Agents.   177
Halocarbon Agents   177
NOAEL and LOAEL and the Need for Precise Design   178
Quantities of Clean Agents   178
Halon Systems and Halon Phase Out   181
Use of Halon 1301   181
Need for Precise Design   181
Alarm Systems   182
Rapid Discharge Rate   182
Storage of Halon.   183
Control Switches for Storage   183
Placement of Halon Cylinders   183
Storage Conditions   183
Estimating Approximate Amounts   184
Partial Total Flooding Halon 1301 Systems   184
Dry Chemical Systems   185

Classification of Dry Chemicals 185
Safety to Personnel 186
Dry Chemical System Design. 186
Tests 186
Local Application Dry Chemical Systems 187
Dry Chemical Hand Hose Systems. 187
Pre-Engineered Dry Chemical Systems 188
References. 188

13 Regulatory Agencies, Authorities and Organizations 189
Regulatory Agencies and Authorities 189
National Fire Protection Association 189
National Electric Code 191
FM Global. 191
Building Codes 192
Types of Buildings: General 194
New Versus Existing Buildings. 194
Fire Marshal and Fire Prevention Bureau. 195
Other Authorities Having Jurisdiction 196
Water Department 196
Summary. 197

14 Fire Suppression System Specifications 199
Construction Specification Standards 199
References. 200
Applicable Standards 200
Design Criteria 201
Hydraulic Calculation Specifications 202
Acceptable Manufacturers 202
Equipment and Material 203
Sprinkler Heads. 203
Testing. 205
Summary. 205
Reference 205

Unwanted Fire and Fire Growth

What is Fire?

We can't discuss manual and automatic fire suppression systems without first discussing fire. This chapter answers the question "what is fire?" and how does it grow. This it contrasts with the actions of occupants and the fire service. Fire is defined by Quintiere (2017) as a "chemical reaction that involves the evolution of light and energy in sufficient amounts to be perceptible". The concept of flashover and its importance to any building fire are also examined.

Fire Growth

Figure 1.1, courtesy of the Home Fire Sprinkler Coalition, shows a fire timeline, in this case for a residential occupancy.

The type of fire we are of course, concerned with are unwanted fires as shown in the Figure. The growing fire is shown in the figure starting as the orange line that transitions to purple at time 1 min 30 s. The fire is seen to transition to flashover at about 3–5 min. Flashover is defined by the NFPA 921 2017 (Guide for Fire and Explosion Investigations) as:

> A transition phase in the development of a compartment fire in which surfaces exposed to thermal radiation reach ignition temperature more or less simultaneously and fire spreads rapidly throughout the space, resulting in full room involvement (FRI)

or total involvement of the compartment or enclosed space.

Once a room has flashed over, it will serve as a source of smoke and heat to ignite surrounding rooms and move smoke throughout a structure. While a single room fire is relatively simple for a fire department to extinguish, multiple room fires become more, sometimes far more complex.

Large fuel sources such as open office plans do not have barriers to contain the fire, but large fires can quickly result from "spreadover".

Fire in the Room of Origin

A determination of the proper detection or suppression device to use depends on a basic knowledge of the function of the different systems, the hazard characteristics, and the stages of fire growth. For the purpose of discussion, the stages of development are summarized in Table 1.1 adopted from Fitzgerald and Meacham (2017).

Stages of Fire Growth

The table refers to the following stages of fire growth:

1. Preburning
2. Initial Burning
3. Vigorous Burning

© Springer International Publishing AG, part of Springer Nature 2019
R. C. Till, J. W. Coon, *Fire Protection*, https://doi.org/10.1007/978-3-319-90844-1_1

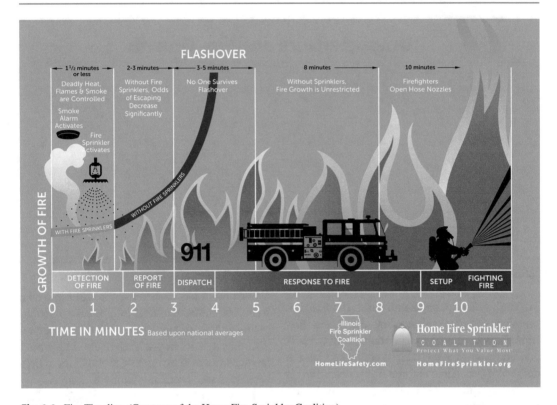

Fig. 1.1 Fire Timeline. (Courtesy of the Home Fire Sprinkler Coalition)

Table 1.1 Fire Timeline

Factor	Preburning (Smoldering)	Initial Burning	Vigorous Burning	Interactive Burning	Remote Burning	Full Room Involvement
Fuel	Surface area receiving heat flux Ignitability	Continuity Ignitability Surface Roughness Thermal inertia of fuel Thickness Surface Area	Continuity and Feedback Thermal inertia of fuel Surface Area Ignitability Quantity	Arrangement and feedback Surface area Tallness Quantity	Quantity Arrangement	
Room Container			Proximity of flames to walls Room Insulation	Proximity of flames to walls Room Insulation Ceiling Height	Room insulation Ceiling height L/W ratio	
Ventilation				Size and location of openings HVAC operation	Size and location of openings HVAC operation	
Overheat Point		Ignition Point	Radiation Point	Enclosure Point	Ceiling Point	FRI

4. Interactive Burning
5. Remote Burning
6. Full Room Involvement (FRI)

Preburning refers to the time from when overheat point to ignition point. It is important in the fire detection realm as certain detectors are able to sense it via the gases that may be produced, and any smoldering. This may go on for a long period.

These four stages from preburning to initial burning are:

1. The incipient stage, where invisible products of combustion are produced. At this stage, smoke is not visible, nor has the heat of combustion developed.
2. The smoldering stage occurs when the combustion has developed to the point where combustion products are visible as smoke, but flame and heat are not yet a factor.
3. The flame stage is an intermediate stage where considerable heat is not yet being produced by the combustion, although this stage very quickly moves into the next, the heat stage. In this stage there may be a clean (laminar) flame, but little or no smoke.
4. The heat stage combines all the elements of combustion invisible products as well as visible smoke, flame, and heat.

Initial burning is the time when the first small flames grow into established burning (EB) – about 19 Btu/s (20 kW) – also referred to as the radiation point.

Vigorous burning is the period of time when a fire moves from established burning until it is about as high as a person (379 Btu/s – 400 kW). It should be noted that once a fire is as high as a person's head, a non-firefighter should consider leaving the area and abandoning any attempt to use a fire extinguisher. The fire is generally developing very rapidly at this point, and an unprotected individual would be in great danger, particularly if they are untrained in the use of fire extinguishers. It should also be noted that in this realm in which heat detectors, including sprinklers may activate – depending on the height of the room and the location of the detector.

Interactive burning is a time in fire growth where flames are beginning to touch the ceiling. The heat release rate when flames touch the ceiling is generally 800 kW to 1 MW and flames will just be beginning to roll over on the ceiling when the fire moves from the Enclosure Point to the ceiling point.

The flame has now mushroomed across the ceiling and some areas remote from the initial ignition may experience auto-ignition. The room moves into flashover.

Finally, the fire moves to full room involvement. This will occur due to flashover or spreadover.

This entire process (up to flashover) is shown as the fire growth "curve" in Fig. 1.1.

The potential hazards should be thoroughly evaluated in order to determine the anticipated combustion stages and how quickly a fire will pass through the various stages. This information is integral to the selection of the proper and most efficient detector/suppression for the hazard area. A fire that has achieved established burning, as shown in the table above, can them grow through up to a point of full room involvement (FRI). One of the main benefits of suppression systems is that, if they are designed properly, the fire will be controlled, and FRI will not occur.

Full Room Involvement

Human beings will not survive flashover. Even in full firefighter turnout gear, humans will be subject to fatal levels of heat and radiation.

When a compartment has flashed over, the fire has changed from a fuel-controlled fire to a ventilation-controlled fire. This means that any air entering the compartment at this point will allow an equal amount of smoke (by mass) to leave the compartment. The quantity of smoke produced post flashover can be extremely large depending on the remaining contents in the room, and the amount of ventilation. It is generally very sooty due to the lack of oxygen at the point of

Table 1.2 Pre and post flashover

	Pre-flashover	Post flashover
Mass of unburned fuel through door	Low	High
Heat release rate in room	Regulated by fuel	Regulated by ventilation
Heat release of door jet	None	High

burning, and it is known that in most fires fatalities are due to smoke, not heat. Post flashover fires are known to produce smoke on the order of lb/s (kg/s).

When a fire is ventilation controlled, the air entering the compartment through whatever openings are provided will limit all combustion occurring inside the compartment. At the same time, there will be combustion outside the compartment of pyrolized fuel that was hot enough to burn inside the compartment, but that did not come in contact with any oxygen until it left the compartment though a vent. This is why "tongues of flame" emerge from windows and doors after flashover has occurred. It should be noted that these flames will also roll across the ceilings of adjacent rooms inside the structure and will likely result in a multi-room fire.

So it has been established that a pre-flashover compartment is quite a different environment from a post-flashover compartment (Table 1.2).

Once FRI has occurred, unburned hot gasses will move out of the room of origin and have the potential to spread from any openings to surrounding rooms. It is much more difficult for a fire the department to extinguish a fire once it has moved beyond the room of origin. Fixed suppression systems can prevent this from happening, and therefore may prevent the destruction of an entire structure.

Notification of Occupants

In this case, one hopes that if the occupants are sleeping, they will wake up upon activation of a smoke alarm and leave the premise. This isn't always the case however, and a portion of occupants perish even in the presence of a working smoke alarm.

Notification of the Fire Department

Notification of the presence of fire, particularly the fire department may be non-trivial. In the case of a protected premise fire alarm system, there may be a local fire alarm, but the fire department may not be notified until someone alerts them. In the era of the cell phone this can be simpler than it was in the past, but it is often assumed by the public that if a fire alarm system is in alarm, the fire department has been notified.

Report of the fire, dispatch and the fire response itself – will likely occur in similar time periods to the Fig. 1.1 graphic. For these cases and others, without sprinklers or some other automatic fire suppression system fire growth is unconstricted.

The level of sophistication of the fire alarm system may be such that finding the fire once the fire department has arrived is a trivial task, but in some cases where no detection or limited detection is present, or the sophisticated fire alarm system is experiencing some sort of fault, the fire may also need to be located by the fire department.

Influence of Fire on Humans

Fire is hazardous to humans for a number of reasons, some obvious and some less so. The more obvious hazards to humans are the physical burns that can happen to people exposed to fire. Less obvious are the exposure to narcotic gasses, irritant gasses, and the limitations on visibility that can influence safe egress. The narcotic gasses include Carbon Monoxide, Carbon Dioxide as well as Hydrogen Cyanide. These gases are responsible for the majority of fire fatalities. All can cause death at different concentrations. The early detection of these gases can activate alarms or suppression systems that can both warn occupants of danger and suppress a fire to make it less dangerous to occupants.

Modeling Fire: Empirically and Numerically

Fire Triangle/Tetrahedron and Suppression

The concept of the fire triangle was developed to provide a simple model of fire suppression. The fire triangle has three sides, the elimination of any one of which will cause a fire to be extinguished. The three sides are Heat, Oxygen and Fuel.

The fire tetrahedron contains the three sides of the fire triangle with an added fourth "dimension". The fourth dimension of the fire tetrahedron is chemical chain reactions. It was added to explain the supression characteristics of some powdered suppressants such as sodium bicarbonate and Purple K, as well as the characteristics of halon gas, all discussed in later chapters.

Classes of Fire: What Is Burning?

Fire can be divided into different classes. This structure helps identify what means can be used to suppress the fire.

- Class A combustibles are ordinary combustibles such as wood, paper, cloth, rubber, and some plastics. These are basically the types of fires we encounter in most buildings.
- Class B combustibles are flammable liquids, paints, grease, solvents, etc. These types of fires are generally encountered in commercial situations.
- Class C is not combustible, but indicates that the agent most be electrically nonconductive, and can therefore be safely discharged on energized electrical equipment, devices, and wiring.
- Class D refers to combustible metals such as magnesium, sodium, and potassium. Agents used for extinguishment of Class D combusti-

bles are not referred to as a dry chemical, but as dry powders.
- Class K – Class K fires are fires in cooking oils and greases such as animal fats and vegetable fats.

Calculation of Flashover

It is possible to calculate the heat release rate at flashover. It should be noted that a Watt is the metric unit of energy commonly used to denote the heat release per unit time used for fires. Specifically, a Watt is a Joule/Second. A Joule is the amount of heat necessary to raise a gram of water about 0.24 (almost a quarter) of a degree Celsius. This small-scale definition isn't particularly helpful to us, as we will generally be concerned about thousands (kilowatts) and even millions (megawatts) when we are discussing fires.

The 2016 NFPA glossary defines flashover as "a stage in the development of a contained fire in which all exposed surfaces reach ignition temperature more or less simultaneously and fire spreads rapidly throughout the space". As discussed later, upper layer temperatures at flashover can be 900–1200 °F(500–600 °C), and the radiative heat flux at the floor is roughly 1.9 Btu/s-ft2 (20 kW/m^2), sufficient to cause paper to ignite.

A simple equation (there are others) for determining the necessary heat release rate (HRR) for flashover was determined by Babrauskas (1980) as:

$$\dot{Q}(Btu/s) = 36.5A\sqrt{H} \qquad (\text{English})$$

$$\dot{Q}(kW) = 750A\sqrt{H} \qquad (\text{Metric})$$

Where A is the area of the opening in sq. ft (sq. m) and the square root of H is the square root of the height of the opening in ft (m). Using this equation, $\dot{Q}$, the HRR at flashover, is roughly 1990 Btu/s (2100 kW) for a room with a standard door, roughly 6.56' (2 m) high and 3.28' (1 m) wide. Note that this calculation applies to a compartment with one opening.

It should also be noted that flashover is NOT the maximum heat release rate that can occur in a compartment fire. An order of magnitude can be noted by comparing an equation for the maximum heat release rate for a compartment with a single opening (Hurley et al. 2015)

$$\dot{Q}\left(Btu\,/\,s\right) = 68.1A\sqrt{H} \qquad \left(\text{English}\right)$$

$$\dot{Q}\left(kW\right) = 1400A\sqrt{H} \qquad \left(\text{Metric}\right)$$

Different configurations may produce different results, but this is the result for a single opening. It can easily be seen that the maximum HRR in a post-flashover compartment can be on the order of double the HRR at flashover.

Fire and Time: Fire Growth for a Single Fuel Package

It is important to understand how to quantify the time it takes for the fire to pass through the phases listed above to achieve flashover or FRI. This is because the time for detector activation, particularly heat detector activation, as used standalone or as the detector in a sprinkler head is a function of the growth rate of the fire (Table 1.3).

The time for a single fuel package to move through to flashover is generally quantified mathematically as an alpha t^2 fire. This is for a single fuel package such as a chair or couch. Each value of alpha corresponds to a different fire growth rate. These rates are given the simple names slow, medium, and fast. Applications of these and their relationship to detector activation is discussed more in the fire alarm section, and the corresponding alphas are shown in Table 1.3.

These types of fires can be "mixed and matched" with the number of fuel packages igniting in the course of the move toward flashover.

Figure 1.2, shows a fire heat release timeline for 5 min. These types of fire growth curves are easily reproduced on spreadsheets and can be used to determine the approximate time to flashover. For example, if the Babrauskas flash over equation determines that a room will flash over when the HRR reaches 2500 kW, a fast fire would flash the room over in a little under 4 min, as

noted on the graph. Examples of how this can be done are located in the next section.

Diagnostic Room Fires

Actual specified fires are generally not part of traditional building codes. A specified fire (a diagnostic or design fire) is usually introduced in "performance-based designs". A method for performance-based design is outlined in the Society of Fire Protection Engineers "SFPE Guide to Performance-Based Fire Protection"(SFPE 2007). A formal introduction of these building design methods will not be adopted in this book; however, the concept of design fire is important.

Multiple design fires may be introduced in order to address the full range of potential issues that could arise. Choosing a relatively large "diagnostic fire" to simulate what the firefighters will encounter has some advantages. A larger fire should produce less visibility, more heat, more smoke, more fatigue, and increased the rate of structural fire attack. These characteristics will make all the buildings' architectural obstacles more difficult than they would be in a smaller fire.

In addition, if the resources to suppress a large fire are brought to bear in an adequate time, those same resources should be able to suppress a small fire in the same rooms of a similar building. This is similar to the concept used in structural building codes for floor loading. If large loads can be handled, the smaller ones can be effectively ignored. In short, a building design that is adequate for large fires should be more than adequate for smaller ones.

Potential disadvantages in the choice of large design fires are: Worst case conditions may not be very useful for determining fire department intervention, as in using this methodology, there

Table 1.3 Alpha and Heat Release Rate

Rate	Alpha (kW/s^2)
Slow	0.002931
Medium	0.01127
Fast	0.04689

Fig. 1.2 Fire growth timeline

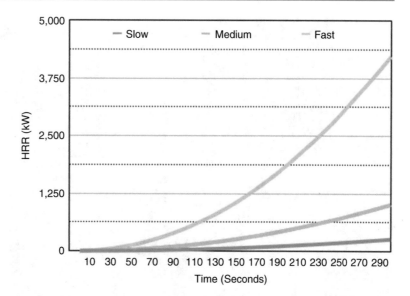

may theoretically be no building left standing when the fire department arrives. If an unreasonably large fire is chosen for the given scenario, the effects of smoke, heat, visibility and potential for structural collapse will be overestimated, and the potential of fire department intervention may be grossly underestimated. This can result in the recommendation of a costly redesign of a proposed structure when none is necessary.

Large design fires have the potential to make detection and location a simple matter when under real conditions this is often not the case. It is very important to address detection and fire location separately.

Office Building Design Fires: Examples

NFPA 72 (The Alarm Code) lists many fuel package fires in its Annex, including heat release rates for furniture and other fuel packages. They are very useful in determining the potential response of sprinklers and other fire detection devices. However, the fires described here incorporate fire in the room of origin but can spread outside the room of origin as well.

Fires must be developed for the associated office building compartments for the type of building layout that is being studied. An associated model for the spread of fire to other compartments

must also be developed. It is important that information about smoke and heat in the burning compartments and their neighbors is available so the influence of these factors on firefighter movement can be established.

For some occupancies such as office buildings, there is a large amount of generalized information available about ventilation and fire loading. Information about the building is vital to the development and integration of the overall model. Fire models such as the zone model CFAST can be used to develop generalized design fires for individual compartment types. The output from these programs includes information on time to compartment flashover and smoke movement information outside the room of origin.

Information about different room types is provided in the work of Culver and Milke (Culver 1976). These studies give sizes ventilation areas and fire loadings for different office building room configurations. These room types are modeled based on their contents and content layout. If these categories can be effectively modeled, then any fire in an office building will be quantified. We will be choosing the more hazardous room layouts, as this is the most significant to firefighters, and measuring time for firefighter accessibility. Design layouts are included for these rooms.

Sizes for traditional offices can be divided many different ways. Figure 1.3 shows the areas

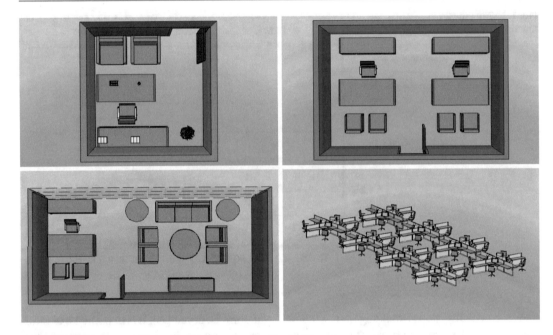

Fig. 1.3 Left to right – small, medium, large and open office spaces

Table 1.4 Office areas, vent sizes and time to flashover

Type	Size	Floor area (Sq. Ft)	Vent area (Medium and slow fire)	Time to flashover
Small office	10′ × 11′	110	80 Sq. ft	150–230 s
Medium office	20′ × 16′	320	160 Sq. ft	160–280 s
Large office	30′ × 16′	480	240 Sq. ft	150–230 s
Open office	2 panel	-	-	~570 s

used for the design offices. This table also shows the vent areas that will be used in the development of medium and slow fires. The vent area for fast fires is taken to be the size of an open door.

For the purposes of demonstration, a set of theoretical design fires was developed for office compartments. In order to construct such a design fire, it is necessary to obtain heat release rates for particular fire packages. This is done so that to flashover for individual rooms, and fire spread throughout buildings, can be modeled.

In developing design fires, we are particularly concerned with time from established burning to flashover. Established burning is defined as the point where the dominant heat transfer mode shifts from convection to radiation. Radiation feedback is assumed to occur when flames reach a height of approximately 10 inches. After this point, the fire grows much faster.

The time to full room involvement is one of the most common calculations performed by fire protection engineers using computer models such as CFAST. There are a number of definitions of FRI including the point where upper layer gas temperatures reach 500–600 °C., or a radiative heat flux at the floor level of 20 W/m². Zone models describe the boundary between the hot upper layer and a cooler lower layer and most account for variations in room ventilation (Table 1.4).

It can be seen that the flashover times for these rooms are on the order of minutes, similar to the time shown in Fig. 1.1.

Open Office Plans and Spreadover

A number of important conclusions were drawn about fires in open office spaces in Harold Nelson's review of the First Interstate Bank Fire of 1988 (Nelson 1989). These conclusions include:

1. The recognition that open arrangements in office settings can develop to flashover. There is a demonstrable fire potential associated with open office arrangements that contain concentrated grouping of combustible work areas. Where such concentrations occur in spaces involving large floor areas and relatively low ceilings, there is usually sufficient combustion air within the space to allow a developing fire to reach flashover conditions. This even if no additional air is introduced into the space. The traditional light hazard expectations associated often associated with offices do not apply in these cases.

2. High space utilization office landscape have the potential, even without the assistance of flashover, of spreading fire over large areas producing fires of major portions.

3. There is an important relationship between the release of fuel from a burning array as the result of heat impinging on it and the availability of oxygen (air.) These relate to efficiency of combustion. Efficiency of combustion is in turn a major determinant of the ability to burn, room layer temperature, carbon monoxide production, oxygen content, fuel transport and flame length.

4. It must be expected that fire products will be spread by natural forces to remote portions of the building given sufficient time. The degree of problem ensuing will be a function of the efficiency of combustion of the fire, the tightness of the shafts and other communicating passages, the presence or absence of smoke control systems, the height of the building, and the weather conditions at the time. Analysis of the potential

involved is important if persons may have to take refuge in the building during the fire.

5. Floor to floor propagation is a potentially serious problem in window wall buildings. The knowledge of flame extension from windows, particularly where petroleum based polymers are involved is insufficient. A better understanding of the relationships between burning rates and flame lengths is needed.

6. In this fire the duration of burning on a floor and the rate of fire propagation from floor to floor were close to each other. A longer duration fire or a faster floor to floor spread could result in an unstoppable fire. Longer duration condition would be expected where a higher total fuel load existed such as commonly occur with merchandising displays or extensive use of combustible interior finishes.

Data is available on workstation heat release rates from a paper by Madrzykowski (1996). Using data he developed design fires were constructed for workstations with 2, 3 and 4 panels respectively. Understanding that "cube farms" can consist of vast areas of open office plan, it is obvious that fire may "spread over" to other cubes very rapidly, and could quickly ignite an entire floor, or even multiple floors with extension outside the building, as it did during the First Interstate Bank and One Meridian Plaza fires. The importance of sprinkler protection becomes particularly important, as these fires can grow very rapidly over the evenly spaced, large fuel packages that open office plans provide (Table 1.5).

It can be seen that open office plans can behave similarly to compartments due to their large heat release rates. Spreadover across these items can result in very large fires that move

Table 1.5 Ceiling heights, number of workstation panels, and time to flashover

Type	Ceiling height	Fire	Time to flashover
Slow fire	9′	2 panel	570 s
Medium fire	8′	3 panel	460 s
Fast fire	8′	4 panel	290 s

through these open areas rapidly. Automatic suppression in the form of sprinklers is one of the few ways of consistently dealing with the problems that these fires represent.

Suppression of the Fire

For the purpose of discussing the fire protection management of buildings in 2018, there are still essentially two kinds of buildings, sprinklered and unsprinklered, and this is also described by Fig. 1.1.

In sprinklered buildings, detection, location and suppression/control are handled by the sprinkler system. A modern functioning wet pipe system will control or extinguish a fire before a room flashes over. In unsprinklered buildings this is not the case, and the fire department and building occupants can be confronted with a far different scenario where smoke and heat are being discharged in copious amounts by the time the fire department arrives, and a much of the building may be just starting to evacuate.

Summary

Figure Fig. 1.1 explains an extremely important concept that this book will repeat – automatic systems can stop fire spread outside of the room of origin immediately and prevent toxic smoke from moving around a premise and impairing or killing occupants who may or may not be aware that the fire is even occurring (for example, if they are sleeping or medically impaired), and destroying property that might otherwise have been saved.

Note that while this Figure demonstrates response for a residential occupancy, it is demonstrated that flashover can occur in the same order of magnitude (2–5 min or so) for other occupancies as well. Fire department setup can change dramatically based on many factors ranging from location of the water supply relative, the weather, the location of the fire and interior conditions. Automatic suppression in the early stages of fire may be the only way of saving a structure and the lives within it.

References

Babrauskas, V. "Estimating Room Flashover Potential." *Fire Technology* 16.2 (1980): 94–103.

Culver, C. *Survey Results for Fire Loads and Live Loads in Office Buildings*. Vol. (NBS Building Science Series Number 85) National Bureau of Standards, 1976.

Fitzgerald, Robert W., and Brian J. Meacham. *Fire Performance Analysis for Buildings*. Wiley, 2017.

Hurley, Morgan J., Daniel T. Gottuk et al. *SFPE Handbook of Fire Protection Engineering*. Springer, 2015.

Madrzykowski, D. "Office Work Station Heat Release Rate Study: Full Scale Vs. Bench Scale." *Interflam '96, 7th International Interflam Conference Proceedings*. Cambridge, England 1996.

Nelson, Harold E. "An Engineering View of the Fire of May 4, 1988 in the First Interstate Bank Building, Los Angeles, California." NISTIR 89–4061 (1989)

Quintiere, James G. *Principles of Fire Behavior, Second Edition*. 2nd ed., New York: Taylor and Francis, 2017.

SFPE. *SFPE Engineering Guide to Performance-Based Fire Protection*. National Fire Protection Association, 2007.

Automatic Sprinkler Heads

2

Early Automatic Suppression Systems

Automatic sprinkler heads were the first practical, widely used, heat detectors. Their history is fascinating.

The earliest piping system intending to serve as a sprinkler system had fittings resembling salt shakers installed in the piping. A cord and weight arrangement controlled the operation of the water supply valve, which was normally kept in the closed position. The water supply was stored in a tank above the piping arrangement, thereby effecting static pressure on the water supply pipe. In the event of a fire, the cord in the fire area burned through, thereby releasing the weights and opening the valve. While the concept was sound, there were several defects. For example, the cords sometimes stretched and lowered the weights, allowing the valve to leak. In other cases, when the system was called upon to function, the valve, (lacking modem gasket and lubricating material) after standing inoperative over a long period of time-would stick and fail to open.

In 1852, the first perforated piping system was installed in America at the Locks and Canals Company in Lowell, Massachusetts. At that time New England was at the center of the textile mill industry. These operations were housed in large, multi-story, wooden structures.

The textile mills contained machinery driven by pulley and belting systems, plus primitive heating and lighting systems which provided a source of ignition. Once ignition occurred, it could readily feed not only on the wooden construction, but also the lint and fabric products inside. Since these buildings were constructed prior to the age of fire protection engineering and modern construction requirements, when a fire did break out and was not immediately extinguished, it would spread rapidly through unprotected openings to other floors. The loss was usually total. The need for an efficient automatic fire extinguishing system became painfully obvious and economically prudent.

Perforated piping systems appeared to be the most practical answer. The customary installation consisted of manually operated control valves serving risers for each floor. Because of this design, however, the water damage was usually severe. For example, a small fire in one corner of a 15,000 or 20,000 square foot floor area meant flooding the entire floor with water. In many cases, the area over the fire might never receive water because perforations would be clogged with paint, sediment, and corrosion.

© Springer International Publishing AG, part of Springer Nature 2019
R. C. Till, J. W. Coon, *Fire Protection*, https://doi.org/10.1007/978-3-319-90844-1_2

History of the Sprinkler: 1864

In an attempt to overcome the problems caused by the clogging of holes drilled directly into the perforated piping system, and to afford better water distribution, the next step in the development of automatic fire suppression systems was the installation of crude, open sprinklers connected to the piping. These open sprinklers were metal bulbs with numerous perforations, which resulted in a spray pattern. Unfortunately, these perforations were also conducive to severe clogging, which not only prevented water discharge, but also disrupted the discharge pattern the head was supposed to develop.

Fusible Element Sprinkler Head

In London, England, in 1864, Major A Stewart Harrison developed the first fusible element sprinkler head. Harrison's device consisted of a hollow brass sphere that contained a large number of counter-sunk holes plugged with solder. When the solder melted from the heat of the fire, the water was discharged. As with previously described methods, the small holes frequently clogged with sediment, making the head at least partially inoperative. In addition, the Harrison head had a rubber valve which had a very short life, and once inoperative, was useless.

There is no record of a patent in Major Harrison's name for this device, and apparently it was never installed in a system. Perhaps this was because of the defects that appeared in the experimental models. Nevertheless, Harrison's sprinkler represents the first automatic application of water from an individual head. But certainly the reports of this idea influenced future sprinkler head development.

Parmelee Sprinkler Head

It is impossible to say just how much influence the Harrison head concept had on the considerable amount of sprinkler head experimentation that was performed from 1874 to 1878. It was during this period that Henry S. Parmelee invented and developed the forerunner of what we know today as the automatic sprinkler head. The Parmelee Number 3, was an upright sprinkler first used in 1875. It consisted of a perforated water distributor completely enclosed in a brass cap which was held in place by solder. When the solder softened from the heat of fire, the cap was forced off by the water pressure, and the water discharged on the fire area.

Parmelee developed this first automatic sprinkler system to protect his New Haven, Connecticut piano factory, and by so doing demonstrated the first practical automatic fire protection system. That same year, Parmelee designed a new model, further improving it in 1878 (Fig. 2.1). In the following years, many improvements resulted in a mass-produced supply.

Skepticism

New inventions are always greeted by suspicion. Parmelee's invention was no exception. Would it work after standing idle for years? Wouldn't water damage be severe if the head operated because of a small fire? If the head or the piping leaked, would the system be capable of controlling a fire of any proportion? Was the expense of installation justified? These questions that were asked in 1878 are still being asked of the systems that are available today. But between 1878 and 1882, the Providence Steam and Gas Pipe Company sold and installed over 200,000 of these crude, but remarkably efficient, sprinkler heads – a testimony to their performance record.

By this time, Factory Mutual Insurance Company (FM) had compiled records clearly indicating the proficiency of automatic sprinkler protection in the facilities they insured. From 1877 to 1888, their records indicated a total fire loss in unsprinklered facilities of $5,707,000, the result of 759 fires, or $7500 per fire. By comparison, there had been 206 fires in sprinkled properties during their 10-year span, with a total dollar loss of $224,480, or $1080 per fire. These statistics convinced insurance companies of the dependability of automatic sprinkler protection.

The companies, in turn, encouraged their policy-holders to install automatic sprinkler systems, thereby protecting themselves against serious fire loss claims. The cost of insurance was reduced if the client installed automatic protection. One example of this sprinkler installation incentive comes from Factory Mutual records which indicate that in 1875, the cost of fire insurance was 30 cents per $100 valuation; in 30 years, this rate had dropped to 4 or 5 cents per $100 valuation.

The Influence of Frederick Grinnell

If one man's name is to be remembered as the father of the automatic sprinkler, it should be Frederick Grinnell. Grinnell was with the Providence Steam and Gas Pipe Company during the Parmelee years. He went on to become President of its successor, the General Fire Extinguisher Company, later named the Grinnell Company – a fire protection company that holds a prominent place in the fire protection industry.

Frederick Grinnell was not only the inventor who perfected the Grinnell sprinkler head, but also a man who could see the life safety and property protection value of automatic protection. He used his genius as both an inventor and a businessman to overcome the public's apathy and skepticism, to render automatic sprinkler installations commercially feasible as well as practical.

Between 1873 and 1882 several sprinkler head inventions appeared which influenced the thinking of practical men like Frederick Grinnell. A sprinkler head developed by Charles E. Buell of New Haven, Connecticut in 1873, incorporated a new feature which was to become a standard design feature on all future sprinkler heads. This head used the principle of a baffle, or deflector, to break up the water, effecting both a greater water distribution and smaller water droplets for greater heat absorption. Between 1872 and 1914, over 450 automatic sprinkler heads were registered with the U.S. patent office, and very likely an equal number were invented but never patented. However, in 1914, Factory Mutual's approval list of sprinkler heads carried only 10 brands, and the 1974 approval list indicated only 15. In 2018, there are now 20 brands.

Despite the numerous advantages offered by fire sprinklers, the high cost of retrofitting fire sprinklers in existing buildings and the lack of integrity in system interfaces are some of the important factors restraining the growth of the fire sprinkler market. The increasing fire protection expenditure of several enterprises to avoid the loss of life and property caused by fire, various norms laid by government agencies, increasing trend of automation in commercial buildings and homes in developing nations, and decreased insurance expenditure are some of the factors driving the market for fire sprinklers.

(Today) the major players in global fire sprinkler market include Tyco (Switzerland), API Group, Inc. (U.S.), Honeywell International, Inc.

(U.S.), Johnson Controls, Inc. (U.S.), United Technologies Corporation (U.S.), Hochiki Corporation (Japan), Robert Bosch GmbH (Germany), Siemens AG (Germany), Minimax GmbH & Co. KG (Germany), and VT MAK (U.S.). (Markets 2015)

The Modern Spray Sprinkler Head further explains the modern spray sprinkler head and examines some of these devices. Spot detection appeared about the same time as sprinklers. The first detectors used roughly the same technology – fusible material that closed or opened a circuit as opposed to a sprinkler head.

The Modern Spray Sprinkler Head

Development of the Spray Sprinkler

Once practical and efficient sprinkler heads were developed, tested, and approved by a nationally recognized testing laboratory, activity was concentrated on ways to increase the cost effectiveness of sprinkler protection. Research toward developing more economical manufacturing methods, less expensive materials, better quality control, and ease of testing has taken place.

Sprinkler Heads

Many companies manufacture sprinkler heads that have been tested and approved by nationally recognized testing laboratories, and each manufacturer has incorporated slightly different features to individualize their own particular sprinkler head. However, the devices are all similar in several respects.

First, all approved sprinkler heads are constructed so that the mechanical pressure on the valve cap far exceeds any anticipated fluid pressure exerted on the bottom of the cap. This feature is incorporated into all sprinkler heads to prevent any possibility of a severe pressure surge, water hammer, or exceptionally high-water pres-

sure and the leakage around the valve cap that may result.

Another feature common to all sprinkler heads is the stamp required on the metal piece of the fusible element stating the temperature rating of the head. In addition to this, every head will have the use-position stamped into the deflector, frame, or fusible link. The manufacturer's name and model number are cast into the frame, and all approved heads will indicate the date of manufacture stamped into the fusible element.

After 2001 a SIN (sprinkler identification number) has been used to identify the type of head. This is located on the deflector. When used with a corresponding technical bulletin, the SIN is used to verify most of the characteristics of the sprinklers described below, in addition to a sprinklers manufacturer.

The frame of most standard sprinkler heads is made of high-grade bronze, and the valve cap or disc covering the water entrance orifice is usually constructed of copper or bronze, or a combination of these two metals. One manufacturer features a copper disc in the cap which forms an expansion chamber between itself and the bronze cap. In this way, if the water should freeze, thereby causing expansion pressure, the copper disc would be punctured by the pressure, and a warning drip of water rather than a complete water discharge would occur to indicate the frozen condition.

Component: The Deflector

As shown in Fig. 2.2 the deflector is the part of the device against which the water stream is directed. The deflector is usually attached to the frame of the sprinkler head by a deflector screw. This screw not only holds the deflector in place on the frame, but also creates a positive tension on the fusible elements of the head. The pressure is transmitted to the cap, which covers the water discharge orifice, thereby

Fig. 2.2 Key sprinkler head components. (Courtesy of John Morrison)

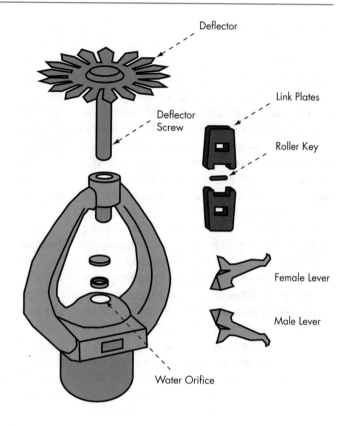

Deflector

Deflector Screw

Link Plates

Roller Key

Female Lever

Male Lever

Water Orifice

keeping the water from discharging. The deflector performs two functions: first, it breaks up the solid water discharge after it leaves the opening, or orifice, into relatively small droplets, and second, it creates the umbrella pattern common to modern standard spray sprinkler heads. This discharge pattern, and water distribution within this pattern, is relatively uniform at all levels below the sprinkler head.

At a distance of 4′ below the spray sprinkler deflector, with the head discharge at 15 gpm (7 psi), the standard sprinkler head covers a circular area of approximately 16′ in diameter. With this uniform pattern of water discharge, fire control and extinguishment is greatly enhanced at the floor level, thereby decreasing the exposure of the ceiling to heat from the combustion.

The spray sprinkler head may be used in the upright or the pendent position, but their positions cannot be interchanged. The Spray Sprinkler Upright (SSU) must be used in the upright position, and the Spray Sprinkler Pendent (SSP) must be used in the pendent position, due to the design of the deflector-large on the SSU and small on the SSP.

Components: The Operating Element(s)

It is in the operating elements of the spray sprinkler head that manufacturers express their individuality. The operating elements can be generally classified as solder-link and frangible-bulb.

The Solder Link

The solder-link sprinkler consists of a combination of levers, links, and struts arranged so that the water or air pressure acting upward on the cap covering the orifice, and the deflector screw force acting in a downward direction, minimize tension in the solder element. By minimizing the force that acts on the solder, a lesser amount of solder can be used, thereby reducing the level of heat necessary to raise the solder to a temperature at which it becomes soft enough to release the linkage. The solder used in sprinkler heads is usually an alloy of tin, lead, cadmium, and bismuth, and is prepared to provide a predetermined, sharply defined melting point. Mixing two or more metals that have different individual melting points can produce a eutectic alloy, an alloy with a melting point lower than the lowest melting point of any of the metals comprising the alloy.

The solder, or alloy, in a sprinkler head has a predetermined melting point that establishes the temperature rating of the specific head. This temperature rating is stamped on the soldered link.

Frangible Bulb Sprinkler Head

The frangible bulb sprinkler is similar in construction to the solder-link sprinkler, except for the operating mechanism, which does not depend on the heating and melting of a solder element. The frangible-bulb sprinkler operating mechanism consists of a special glass bulb which takes the place of the mechanical linkage in the solder-type head but performs the same function of holding the valve cap over the orifice against the water or air pressure. This glass bulb contains a liquid with high expansion qualities. However, the liquid does not completely fill the bulb, allowing for a small air bubble. A pendant frangible head sprinkler is shown next to a fusible element sprinkler head in Fig. 2.3.

When the heat from the fire reaches the head, the liquid expands, compressing the air bubble. If the heat continues to be absorbed by the sealed glass bulb, the liquid continues to expand until it has completely absorbed the air bubble, and the bulb becomes completely filled with liquid. With the cushion of air gone, and the liquid continuing to expand, the pressure inside the bulb increases until it shatters-literally explodes-thereby releasing the valve cap and allowing the water to discharge against the deflector. The air bubble size is a very important design feature of this head, because by adjusting and regulating the ratio of liquid to air when the bulb is sealed, the exact operating temperature of the head can be determined.

Temperature Rating

Table 2.1 indicates the temperature rating and corresponding color coding of solder-link sprinkler heads used in areas where the maximum ceiling temperature has been determined. If the range of room temperature is not known, or if a piece of equipment in the area will produce heat above the ambient room temperature, it may be necessary to hang a recording thermometer in the area in question to determine temperature conditions. The recording thermometer records the highest temperature attained. Unlike ordinary thermometers, it will remain at this temperature until reset.

The color code indicated in the table is painted on the sprinkler head frame so that an inspector or maintenance worker can identify the temperature rating of the head without having to examine its stamp. The use of color on the head to identify a fusing temperature above the ordinary classification is not required for decorative heads, such as those installed below a suspended ceiling, or on chrome-plated heads. Unless otherwise specified or indicated on a plan, the temperature rating of an ordinary sprinkler head is 165 °F.

K-Factor

The k factor stamped on every sprinkler determines the relationship between the water pressure behind a sprinkler head and the amount of flow that the sprinkler head will produce. This relationship is shown below.

Fig. 2.3 Pendant SSP
sprinkler heads.
(Courtesy of Reliable
Sprinkler)

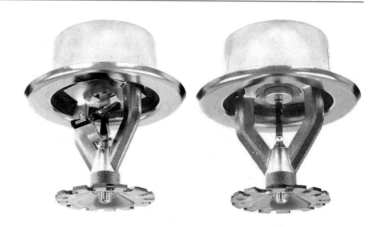

Table 2.1 Temperature ratings, classifications and color codes

Temp. Rating °F,	135–170	175–225	250–300	325–375	400–475	500–575
Temp. Class	Ordinary	Intermediate	High	Extra high	Very extra high	Ultra high
Glass Bulb Color	Orange or Red	Yellow or Green	Blue	Purple	Black	Black
Color Code	Uncolored or Black	White	Blue	Red	Green	Orange

$$Q = k\sqrt{P}$$

As the suppression requirement of certain hazards (e.g. plastic storage) has continued to rise, so has the K factor. A typical K factor is 5.6 (1/2″ thread), but in some storage occupancies they can get as high as 32 as of this printing (1 1/4″ thread).

The k factor becomes an important element as the fire size at activation gets larger. Clearly more water is required when a larger design fire is expected. The k factor may be raised accordingly to accommodate this.

Development of the Large Orifice Head

The new spray sprinkler was a solution to the threat of the high-challenge fire posed by industrial advancement. It was also an answer to better fire protection at a smaller investment. Nevertheless, the interest in improved fire protection was not dulled by this radical change in sprinkler head design. High-level fire challenges-fires that were beyond the capabilities and discharge capacities of standard spray sprinkler installations continue to drive the development of new sprinklers.

When an intense fire develops, it can create exceptionally strong fire plumes. As a result, the discharge from standard sprinklers operating over such a fire may be unable to penetrate the fire plume, with the droplets turning to steam far above the burning fuel surface, or combustion core. With the expanded spacing that the spray sprinkler allows, an excessive number of sprinklers around the perimeter of the fire center will begin to open when the first sprinklers directly over the fire fail to control the combustion. As more and more heads open, the discharge density of the heads directly over the fire will decrease, producing less efficient fire control.

Modern technology demanded a sprinkler head with a higher water discharge and density. It also required sprinklers with droplets in the discharge large enough to penetrate the kinds of fire plumes found in intense fires from combustible products that were not anticipated when the spray head was first installed. The improved sprinkler head would also have to cope with fire plumes resulting from storage configurations of a height and extensive square footage area unheard of before. Development of the large orifice head was the result: a sprinkler head similar to the standard sprinkler with a design exception a 17/32″ orifice.

It was discovered that an orifice 1/32″ larger than the normal 1/2″ orifice would deliver 140% more water, at the same pressure. This proved exceedingly advantageous when there was a demand for high water density without the corresponding demand for high water pressure to meet the density requirements. The fire protection engineer could now meet the high density gpm per square foot requirements of rack storage, plastic and rubber storage, flammable liquid operations, and a multitude of situations occasioned by new products and processes and changing industrial storage and operational demands. These needs could be met with smaller pipe sizes and less pressure than the orifice head required. The chapter Hydraulic Calculations of Sprinkler Systems, describes in detail the importance of demand and supply conditions, and points out the advantages of the large orifice head over the regular size in satisfying unusual demand and supply situations.

The pipe thread connection of a large orifice sprinkler is 3/4″ NPT (National Pipe Thread). The standard orifice sprinkler has a NPT pipe thread connection. This explains why the large orifice head is sometimes referred to as the 3/4″ head.

For some specialty situations, the orifice can be the 17/32″ standard, but the pipe thread connection is only 1/2″ NPT. With the proper hydraulic calculations and complete re-evaluation of the system layout, it is possible to remove the standard spray sprinkler (orifice) from an existing system and to install this head in its place.

Small Orifice Head

While the large orifice head is used where a high density (gpm per square foot) is required and/or the residual pressure (the pressure left over which shows on the gauge, when water is flowing) is low, the small orifice head is used when low density is required, and/or a high residual pressure is available. Smaller orifice heads discharge less water than the standard orifice head at the same pressure. They are used in both light and ordinary hazard areas where there is a limited water supply, for conservation and efficient limitation of water. The application of the small orifice head is limited and must be recommended with great care by experienced fire protection engineers, and only after all of the protection circumstances have been carefully evaluated. The small orifice head is available in several orifice sizes: 1/4″, 5/16″, 3/8″, and 7/16″. When considering the use of the small orifice head, one should keep in mind that increasing the orifice head by 1/32″ increases the discharge by 40%. For an example, at 65 psi at the sprinkler head, the orifice head will discharge about 45 gpm, the 17/32″ orifice head will discharge about 66 gpm, and the small orifice 1/4″ head will discharge about 13 gpm.

To avoid mistaking a small orifice head for a standard head, every small orifice head with an orifice less than has a very noticeable pintle, or extension of the compression screw, which extends to above the deflector. In this way, an inspector can tell at a glance which size sprinkler head he or she is dealing with.

Dry Pendant Sprinklers

When the sprinkler system piping is installed above a suspended ceiling (concealed piping), and the sprinkler heads are all that show below the ceiling (down feed), the drop nipple from the sprinkler line to supply these pendant heads below the ceiling will obviously fill with water when a dry system trips and will not drain when the dry pipe system is emptied of water. The industry solved this problem with the development of the dry pendent sprinkler.

A dry pipe system is installed in areas subject to freezing. Water is only admitted to the system piping when a head fuses and the compressed air that holds the water out of the piping is exhausted through the open head.

Once a dry system has tripped, or filled with water, the open head or heads must be replaced, and the system must be drained of all water and refilled with compressed air to prevent the water in the pipes from freezing. A review of a cross-section of a sprinkler head will reveal a pocket from the valve cap on the orifice to the end of the male pipe thread where the head screws into the pipe fitting, and where water could collect if the head was installed in the pendent position. This water would remain when the dry system piping was drained and would be subject to freezing.

A dry pendent sprinkler head consists of a drop nipple with a standard pendent head sprinkler attached. Inside the drop nipple is a specially designed shaft that holds a cap at the top of the nipple to prevent water from entering the drop nipple. This shaft is under tension in order to hold the cap against the water pressure. This tension is maintained by engaging the shaft with the operating elements of the sprinkler head. When the sprinkler head fuses and the operating elements are released, the restraint on the shafts is relieved. It then drops a predetermined distance, which pulls the cap away from the opening, allowing water to enter the nipple, flow around the sides of the shaft, and discharge through the orifice of the head.

Unlike the standard pendent sprinkler heads, the dry pendent sprinkler cannot be cut-to-fit in the field due to the single unit construction of the sprinkler head and drop nipple. After the concealed line piping of a dry pipe system is installed, accurate measurements must be taken from the face of the fitting (the bead of the line tee outlet) to the underside of the suspended ceiling. These dimensions are then used to order the dry pendent sprinkler heads (Fig. 2.4).

Note that when the dry pendent head is screwed into place, it projects a considerable distance into the fitting. For this reason, the dry pendent sprinkler must always be connected to a tee and never to an elbow. If used in an elbow, the sprinkler head could impinge on the curvature of the fitting and could also create turbulence that could affect the flow of water from the fitting to the sprinkler head. Therefore, when dry pendent heads are installed on the end of a sprinkler line, the elbow must be replaced with a tee and nipple and cap, with the dry pendent sprinkler installed in the tee outlet.

Dry pendent heads can also be used on wet pipe systems to protect a freezer or cooler. For example, a supermarket may be protected with a wet pipe sprinkler system because the store is heated and not subject to freezing. The insurance carrier, however, might require protection of a small freezer box where a pendent sprinkler supplied from the wet pipe system would surely freeze. A dry pipe sprinkler in the freezer box supplied by the wet pipe sprinkler piping would be appropriate in this case.

In this example, if the top of the freezer box were several feet below the ceiling where the wet pipe system piping was installed, it would be necessary to drop down from the ceiling with the supply pipe for the dry pendent head and install the horizontal tee and nipple and cap for the dry pendent head a few inches above the top of the freezer box. Dry pendent heads are expensive, and the cost of the head is based on the length of the dry pendent head unit. The same design criterion is true for dry pendent heads on a dry pipe system.

Sidewall Sprinklers

Sidewall sprinklers are similar to standard sprinkler heads, with the exception of the deflector. The deflector produces a discharge that looks somewhat like half of a normal sprinkler umbrella pattern. Installed on one wall of a room, a sidewall head will distribute, or throw, this pattern across the length of the room, covering the entire square footage of the room.

Fig. 2.4 Dry pendant sprinkler head. (Courtesy of Reliable Sprinkler)

Fig. 2.5 Sidewall
sprinkler heads.
(Courtesy of Reliable
Sprinkler)

Naturally there are limits to the discharge distance of regular sidewall sprinklers, and they should be installed according to their listing.

There are currently on the market sidewall heads called extended coverage heads that are designed for light hazard areas where extra-long water spray patterns are needed; for example, in above-average sized rooms in hotels, nursing homes, motels, and so forth where standard sidewall heads are impractical (Fig. 2.5).

The extended discharge sidewall head patterns are mentioned here for reference only, because their installation and use must be overseen by experienced fire protection engineers and sprinkler designers. Sidewall sprinklers are available in the upright, pendent position, as well as in the horizontal positions. They are installed along sidewalls, where their deflector discharges the majority of the water away from the wall, with only a small portion discharged against the wall behind the sidewall head. Upright and pendent sidewall sprinkler deflectors should be positioned at a distance from walls and ceilings not more than 6′ 10″ or less than 4 10″. Horizontal sidewall sprinklers are allowed to be positioned less than 4″ from the wall, and 6″ to 12″ below noncombustible ceilings. The following example demonstrates the advantages of sidewall sprinklers. Installing a sprinkler system in an existing hotel where the ceilings are plastered and installing piping above the ceilings with pendent heads would require tearing out the plastered ceilings and replacing them. By using sidewall heads and installing the sprinkler piping close to the ceiling and close to one wall of the rooms where it will not be noticeable, the rooms can be protected without disturbing the existing ceilings. This is but one of hundreds of examples in new and existing construction where sidewall heads can

provide adequate protection. However, these special discharge heads must only be used under the recommendation of an experienced engineer, so as not to stretch their protection parameters beyond their limitations.

Decorative Sprinklers

The sprinkler industry has responded to the restrictions by architects that regular pendent sprinklers showing below their suspended ceilings are too unsightly. The result-the industry has provided a vast array of decorative sprinklers.

In designing a more attractive device, the sprinkler industry knew that as long as the deflector of the pendent head was one inch below the ceiling when the head discharged, they could continue to provide a suitable discharge pattern. With this in mind, they proceeded to conceal as much of the body of the sprinkler head as possible.

In preparing a specification for the use of a decorative sprinkler head, one must be careful not to use generic terms like flush head. It is imperative to identify the type of head being specified by a manufacturers name and model number-or engineer approved equal. The reason for this is best explained by browsing through the illustrations of decorative heads. Some manufacturers refer to the recessed head -which is just a pendent head recessed in a cup-type ceiling escutcheon-as a flush head.

There are decorative heads available that are completely concealed, where the only part that shows below the ceiling is a round plate. The plate is held in place by a fusible material that will melt before the fusible element of the sprinkler head melts. When this happens, the plate will drop off exposing the sprinkler deflector. The deflector, in turn, drops below the ceiling when the head fuses.

Naturally, decorative heads are more expensive than regular brass sprinkler heads, and the more complicated or the more concealed they are, the more expensive they will be. The use of chrome pendent sprinkler heads, which are slightly more expensive than brass pendent heads, may prove an acceptable alternative to decorative heads as will the recessed pendent head. However, in areas where appearance is extremely important, the additional cost of a truly flush, or totally concealed head may well be worth the additional cost.

Old Style Versus New Style Sprinkler Head

Sprinkler heads manufactured prior to 1953 are now referred to as old style sprinklers because of the drastic industry-wide change that occurred that year with the introduction of the spray sprinkler. The old-style sprinkler had an exceptional record of property protection, such that fire insurance companies were cutting the premiums of sprinklered properties as much as 80%. However, construction methods and materials as well as occupancy patterns were changing, and manufacturing and warehousing areas were increasing in size. All of these factors put a tremendous strain on the sprinkler systems ability to extinguish, or even control, high-intensity, large-area fires.

Sprinkler system research at this time concentrated on developing a standard sprinkler head that could control the excessive fire loading created by modern technology and construction, while meeting the increasing demands of industry. The old-style sprinkler head had discharged approximately 50% of the water upward onto the ceiling, which meant that only 50% of the water discharge from the head landed on the fire in a uniform pattern. The objectives of the new research were to discharge more water on the fire without increasing the water pressure and volume, and to break this water discharge up into smaller droplets, thus exposing more water surface to absorb the heat by evaporating to steam.

The new standard spray sprinkler head still used a 1/2″ orifice discharge outlet similar to the old-style head. In fact, with few exceptions, all the features of the old-style head were similar to the new-except for the deflector, which was completely redesigned to direct approximately 98% of the water downward onto the floor, or fire area. This new type of deflector broke the solid discharge stream through the 1/2″ orifice into a water spray consisting of droplets of water smaller than those released by the old style head discharge. Now it was possible, theoretically, to extinguish or control a fire that was producing twice the amount or intensity of heat because the head could now deliver twice the amount of water directly on the fire in a controlled pattern and density.

Special Heads

The introduction of this spray sprinkler allowed an increase in head spacing (square foot coverage per head) because of the expanded and uniform discharge pattern. This meant fewer heads were required to cover the same area. Smaller pipe sizes and valves could be used, which resulted in less total system cost without sacrificing fire protection efficiency. Now that the sprinkler industry had a sprinkler head that discharged 98% of the water onto the floor, a relatively simple mathematical formula could be used to determine the amount of discharge reaching the floor at various head pressures. Pressure versus discharge is discussed in the chapter on Hydraulic Calculations, but it is important to note here that it was the spray sprinkler that introduced to the sprinkler industry the use of hydraulically calculated sprinkler systems.

There are a number of special heads available today. They are briefly described here for reference purposes only, as they have very limited applications and should only be specified and installed by experienced fire protection designers.

Large Drop Sprinkler

A system for storing commodities is the use of rack storage, with some system racks reaching heights of 100. A fire in a rack storage warehouse produces extremely high heat release and rapid fire development that can easily overpower the heavy discharge of a sprinkler system. The narrow aisles and flue spaces between the racks create a chimney effect, with heat plumes racing up the face of the rack storage at speeds exceeding 30 miles per hour.

Spray sprinklers installed at the ceiling and in the racks themselves produce large volumes of very fine droplets of water. However, tests have proven that the majority of these droplets -up to 98% – are unable to penetrate the high velocity heat plumes. In response, the industry has developed a head called the large drop sprinkler which, as its name implies, produces very large drops in the discharge. Tests with these large drop sprinklers indicated that significantly more of the water discharged penetrated the heat plumes and extinguished the combustion. The balance of the discharge from these large drop heads performed cooling that assured control of the combustion until the fire was suppressed.

Early Suppression Fast Response (ESFR) Sprinkler

Factory Mutual Research combined the advantages of the quick response sprinkler head with the advantages of the large drop sprinkler, and the result was the early suppression fast response. ESFR heads provide the protection that their name implies: They respond much more quickly than regular heads with the same temperature rating and discharge a density with drop sizes that provide fast suppression of combustion. These ESFR heads offer tremendous advantages in suppressing the high-challenge fire, especially rack storage fires, but systems using these heads require a high level of fire protection engineering expertise and should never be specified or recommended by the layman.

On-Off Sprinkler

Another special sprinkler head is the on -off sprinkler. This head automatically resets itself after a fire has been extinguished. Water is discharged when the temperature rating of the head is reached; when heat from the fire drops below a predetermined temperature, the head automatically turns off.

One manufacturer describes the operation of its on-off sprinkler as such: When a bi-metallic snap disc is exposed to heat in excess of 165 °F, it snaps open and allows a small amount of water to pass through a small pilot orifice. This passage of water releases pressure to a piston assembly, forcing the piston down and allowing the water to be discharged through the main orifice or port. When the fire is controlled and the heat decreases, water pressure builds up in the pressure chamber and forces the piston assembly closed. The discharge flow of water is arrested, and the sprinkler head is therefore automatically reset when the snap disc returns to the operational position.

These on-off heads should be used with great care and only with the approval of the buildings insurance carrier. When a standard sprinkler head fuses it remains open until someone in authority determines that the fire has been extinguished and the water supply to the system may be closed and the fused head replaced.

Insurance carriers want the reliability of a standard sprinkler system, as well as a physical assurance that the fire is out. They generally will permit the use of these on-off heads only for a very special condition and even then they will limit the number of on-off heads used to a particular area or hazard.

Corrosion-Resistant Sprinklers

Corrosion-resistant sprinklers also fall under the special head category. These are standard sprinkler heads that have been coated at the factory with wax or lead to prevent the head from corroding. When a sprinkler system is installed in an area containing chemicals, moisture, corro-

Fig. 2.6 Chrome frangible pendant sprinkler head. (Courtesy of Reliable Sprinkler)

sive vapors, or where the heads are exposed to the elements, corrosion-resistant heads must be installed. According to the specifications, the corrosion-resistant coatings shall be applied only by the manufacturer of the sprinkler head. Any homemade coating application will very likely render it an unapproved head. Furthermore, if the protective coating is damaged during installation, it must be repaired using only the coating furnished by the sprinkler manufacturer and applied in an approved manner, so that no part of the head is exposed to the corrosive atmosphere (Fig. 2.6).

Rack Sprinklers

The sprinkler heads used in storage racks also fall under the special head category. These are standard heads with one exception; they have a water shield about three inches in diameter attached above the deflector on upright heads and attached above the frame of the head on pendent heads. The reason for this is simple. These sprinklers are installed in storage racks at various levels of the rack-this is in addition to a sprinkler system at the ceiling of the rack storage warehouse. When a sprinkler at the ceiling discharges, or when a head located higher up in the rack discharges, it will throw water on the heads at lower elevations. This water would keep the heads below the discharge wet and cool so that they cannot absorb sufficient heat to fuse. This phenomenon is called cold soldering. The water shield on these lower heads keeps the water discharge (from above) from wetting and cooling the fusible links of the lower heads. In short, they act as umbrellas for the heads to keep the fusible links dry.

It is important to note here that cold soldering does not occur in rack sprinkler systems alone. Sprinkler heads on any sprinkler system must be installed no closer to one another than 6′ to prevent cold soldering. On standard, horizontal, sprinkler systems, never install sprinkler heads closer together than 6′ or one head discharge may prevent another head from fusing. Since a rack sprinkler with a fixed water shield is more expensive than a regular head without the water shield, there are on the market separate water shields that can be attached in the field to regular upright and pendent sprinkler heads. These separate water shields come with a sprinkler head guard, which is simply a wire enclosure around the head. In rack sprinkler systems, where commodities are constantly being added or removed from the racks, the probability of damaging the head, especially the fusible link, is very high. The head guard protects the rack sprinkler head from physical damage.

CMDA (Standard Spray)

CMDA sprinklers are used to protect both storage and non-storage systems. Non-storage systems include offices, manufacturing, and retail. Some of the options for storage systems provide standard densities such as 0.495/2000 or 0.33/3000. These densities are designed in the worst case to control the fire to within a design area until the fire department arrives and extinguishes the fire.

CMSA (Large Drop)

A type of spray sprinkler that is capable of producing characteristic large water droplets and that is listed for its capability to provide fire control of specific high-challenge fire hazards. Similarly, it also allows for protection of commodities in buildings with higher ceilings without (in most cases) the need for in-rack sprinklers.

These systems like CMDA are only designed to control the fire, not extinguish it. Since the CMSA sprinkler heads are more expensive than a standard CMDA head, and yet there are no "breaks" given as with ESFR (for example smoke and heat vents are still required), these type of sprinkler systems are not used as often as CMDA and ESFR systems.

Sprinkler Accessories

Head guards should be used on any sprinkler system where the heads are subject to physical damage or are within reach. Head guards can also be obtained for some decorative heads.

Every sprinkler installation must have a supply of spare sprinkler heads in every type and temperature rating used in the system. The spare heads are to be installed as quickly as possible so that the system may resume service without delay. The NFPA dictates how many spare heads shall be on hand based on the number of heads in the system. These spare heads are to be stored in a cabinet designed specifically for that purpose. Contained within the cabinet will also be a sprinkler wrench which is designed to fit the heads installed in the particular system. This cabinet is usually installed adjacent to the sprinkler riser.

Other Common Characteristics of Spray Sprinkler

Sprinklers need to be replaced on a 50-year basis (unless a representative sample is tested). Older sprinklers are slowly being phased out. Their use may result in the sprinkler system being over-whelmed by modern plastics. An example of this type of hazard is discussed in Fitzgerald and Meacham (2017).

In the past, a sprinkler head with a temperature rating of 165 °F could operate when the head reaches a temperature of 165 °F, plus or minus 5 °F. However, because of the heat absorption of the mass of the fusible link, the sprinkler head frame, the piping, etc., the air temperature may be as high as 1000 °F before the fusible element operated. This phenomenon can be referred to as thermal lag.

In response, the industry has developed a sprinkler head with an extremely sensitive fusible element that overcomes most of the thermal lag. As a result, a 165 °F head could operate approximately six times as fast as a sprinkler head with a conventional fusible element temperature rated at 165 °F.

To deal with this situation, the concept of Response Time Index (RTI) was developed. RTI is defined by NFPA 13 (2016) as "A numerical value that represents the thermal response sensitivity of the sensing element in a heat detector, sprinkler, or other heat-sensing fire detection device to the fire environment in terms of gas temperature and velocity versus time." Sprinklers defined as fast response have a thermal element with an RTI of 50 (meters-seconds)$^{1/2}$ or less. Standard response sprinklers have an RTI of 80 or more.

The differential between the temperature rating of the head and the ambient temperature affects the speed of operation. A head rated at 212 °F, installed in an area where the surrounding atmosphere is 0 °F, may need to absorb considerable heat to reach operating temperature depending on the RTI of the head.

The number of sprinklers that are anticipated to open in a fire is based on several factors: height of the heads above the floor, square footage each head covers and, most importantly, the fire hazard classification of the occupancy. In addition, regardless of the temperature rating of the heads, it is obvious that the heads directly over the combustion will fuse first, and as the heat from the combustion mushrooms at the ceiling in the form of ceiling jets, the heads around the perimeter of the combustion will also be activated.

The heat from the fire may continue to spread across the ceiling, and although the temperature decreases at the edges, it is still sufficient to activate more heads via ceiling jets. The discharge from heads that are too far from the origin of the fire to contribute to fire suppression, control, or property protection will only cause water damage and tax the water supply to the operating heads. Therefore, if all the heads in the area are temperature-rated only slightly above that required by the ambient temperature of the area, the aforementioned heads on the fringes of the ceiling heat will not fuse as rapidly as heads with a lower temperature rating, preventing unnecessary water damage.

Heads that may fuse too quickly can use up the water supply needed to put out the fire. The chapters on Hydraulic Calculations of Sprinkler Systems, describes how extremely important it is to know how many heads will fuse before the water supply is no longer capable of furnishing sufficient water to the system. It will show the importance of sprinkler temperature rating in preventing heads that are outside of the calculated area from fusing. Establishing the proper temperature rating of sprinkler heads from a chart when the ambient temperature is known, and no unusual heat producing equipment is present, is relatively simple.

It is imperative that the temperature rating of sprinkler heads above the standard requirements be accomplished only by an experienced fire protection professional. Recommending heads with higher temperature ratings than required based on the layman's view of preventing water damage can be very dangerous. These high-temperature heads over the combustion may not fuse while the fire is small and the discharge may come too late to prevent combustion from intensifying and spreading. High-temperature heads must be used under skylights where the heads are exposed to the direct sunlight or located in unventilated attics, in confined spaces under uninsulated roofs,

and in unventilated areas having high-powered electric lights on the ceiling. Specifications usually specifically state that high-temperature heads shall be provided in the area of heat-producing equipment. The specifications also make recommendations as to the location of the head.

The decision to use higher-than-required temperature-rated heads must be made based on experience and a thorough understanding of fire protection. Otherwise, the time delay in fusing a head, and the opening of too few heads, can allow a fire to develop to a point that is beyond the ability of the sprinkler system to control. A fire that develops rapidly and increases ceiling temperatures quickly will operate a 212 °F sprinkler head almost as quickly as a 165 °F, head, but a fire that develops slowly, with a slow rise in ceiling temperature, will introduce a considerable time differential between the activation of the 165 °F, and 212 °F sprinkler head.

Summary

Automatic sprinkler heads were the first practical, widely used, heat detectors. They have a fascinating history and are now available for many different applications, including areas that need to stay dry most of the time, but still have high quantities of fuel, such as refrigeration units. They have the advantage not only of detecting fire (though the use of flow switches and other technology discussed later), but also of immediately controlling and possibly suppressing it.

References

Fitzgerald, Robert W., and Brian J. Meacham. *Fire Performance Analysis for Buildings.* Wiley, 2017.
Markets 2015. <https://www.marketsandmarkets.com/Market-Reports/fire-sprinkler-system-market-3646511.html>.

3

Other Detection and Alarm Devices

Detection and Alarm Systems

Selecting the proper fire detection and alarm systems requires a careful evaluation of all that the system is designed to accomplish as well as the hazard and the environment in which the detection system must operate. A detection system must fulfill two basic requirements. It must be sensitive enough to detect combustion within an acceptable and required time frame, and it must be reliable enough to not respond to non-fire activation sources. A determination of the proper detection system to use depends on a basic knowledge of the function of the different detectors, the hazard characteristics, and the four stages preburning initial burning (Chapter 1). It also must be pointed out that while some systems may be more sensitive to fire and smoke, they may not offer the unique advantage that sprinklers do – water where it is needed and when it is needed.

Alarm Initiating Devices

Alarm initiating devices are hardware components that communicate the presence of fire to the Fire Alarm Control Unit (FACU), and there are many types. Some are activated by people such as manual pull stations, while others automatically activate under other conditions that occur as the fire grows such as increasing levels of smoke and heat.

Manual Pull Stations

Manual pull stations, or as NFPA 72 (2016) refers to them "manual fire alarm boxes", are essentially switches that activate a fire alarm by opening or closing a circuit. They are often referred to as pull stations. The two varieties are single-action pull stations, which only require one step to activate (as shown in Fig. 3.1), or double action pull stations that require two steps. The double action stations help prevent false alarms.

Thermal Detection Systems

Thermal detectors are designed to actuate at a predetermined fixed temperature, and they should be used when the protected hazard area has the potential to produce considerable combustion heat, or at least sufficient combustion heat to activate the detector before considerable damage is done passing through the early stages of the combustion.

A sprinkler head is a thermal detector. When combustion heat reaches the temperature setting of the fusible element of the sprinkler head, the fusible element melts, or softens sufficiently to

© Springer International Publishing AG, part of Springer Nature 2019
R. C. Till, J. W. Coon, *Fire Protection*, https://doi.org/10.1007/978-3-319-90844-1_3

Fig. 3.1 Pull station. (Robert Till)

release the head linkage, thereby releasing the discharge. The water movement in the piping actuates a device that transmits an alarm. The sprinkler head is also a good example of a non-restorable detector. Once the head is activated, it must be replaced.

Some other types of thermal detectors are also non-restorable. A thermal element melts at a predetermined temperature, releasing a mechanical arrangement such as a spring or a plunger which makes or breaks an electric circuit, activating the alarm system. Other thermal detectors are available that activate at a predetermined fixed temperature but are self-restoring. After the detector has been activated and the combustion heat removed, the detector returns to room temperature and automatically resets itself.

One type of self-restoring thermal detector operates on the principle of a disk made of two different metals, with a curved shape. When the detector is subjected to combustion heat that reaches the fixed operating temperature of the detector, the disk curvature is reversed, transforming from concave to convex shape. This reversal of the disk allows a spring tension plunger to drop, which makes or breaks an electric circuit, thereby activating the alarm or suppression system.

One caution should be noted regarding the use of self-restoring detectors. Unless equipped with special devices, some suppression system valves that are operated when a thermal detector is activated can close when the thermal detector cools and is restored to a normal operating condition. When a suppression system has been activated, it is important that the system not be shut down until the combustion is completely extinguished, and in some instances until the surrounding area has cooled sufficiently to prevent a re-ignition of the combustion.

Once the flaming stage of the combustion has been controlled, the temperature at the ceiling could drop below the operating point of the detector and the detector could reset, re-opening or closing a solenoid valve and shutting down the suppression system valve automatically. When designing a detection system that activates a suppression system, and the detectors are self-restoring, always determine what may be required to maintain the suppression system in an operating condition should the detectors reset. Unless a suppression system is designed to shut down automatically, the system should not be shut down until a professional firefighter determines that it is time to close a manual valve.

Pilot Head Systems: Sprinklers Themselves

The pilot head detection system is a thermal detection system frequently used to activate suppression systems. The pilot head system uses regular sprinkler heads as the thermal detectors. These heads are installed on 1/2″ pipe which is filled with compressed air. When a head fuses, the compressed air is released, and the release of compressed air pressure activates the suppression system valve.

The pilot head system is very reliable in that it is not subject to false trips from non-fire sources, which can cause havoc with other detectors that operate on a principle other than heat.

Pilot head detection systems are frequently used to activate cooling tower and outdoor transformer deluge suppression systems. Since these systems use regular sprinkler heads, the heads must be located so that sufficient heat can collect to fuse the head. When the system is located within a structure, the location of the heads should closely follow the rules for sprinkler head location below a ceiling. If the system is used outdoors, as it would be with transformer protection, each pilot head should have a heat collector installed above the head.

Continuous Line Detector

The continuous line detection system, sometimes referred to as a continuous wire detection system or thermistor system, uses a detector with the appearance of a wire. A simple description of one type of line detector is a metal tube containing a wire, or conductor, surrounded by a special compound. When a section of the line detector is subjected to the heat of combustion, the compound loses its electrical resistance and a small current passes from the wire in the core of the tube to the metal tube. The control panel senses this current flow and activates the detection system devices.

The line-type detector is self-restoring if the fire does not physically damage the equipment. Line detectors come in various lengths joined together at junction points. The line detector can indicate on an annunciator panel the exact location of the line where the combustion heat has activated the line. The line detector is also available in different temperature ranges.

The line detector has many advantages when used in long conveyor enclosures and in conveyors without enclosures as well as in and around irregularly shaped equipment. It is also used on the ceiling of a building, similar to the placement of spot-type thermal detectors.

The line-type detector has excellent reliability in terms of not responding to non-fire sources of activation, but like any thermal detector, it is not sensitive enough to be referred to as an early warning detector, since it must have sufficient combustion heat to react.

Protectowire Type Detector

Protectowire type detectors (named after one of the first companies to produce them) consist of two conductors separated by a heat sensitive polymer. The polymer melts at a fixed temperature (135 °F for example) and an alarm is initiated. The location of the fire along the line of fire can be determined based on the resistivity of the metals used. This type of detector is shown in Fig. 3.2. Digital versions of these systems are now able to discriminate between physical damage to the cabling and an actual fire event.

Rate of Rise Detector

The rate of rise detector responds to a rate of temperature rise of 15° per minute regardless of the rooms ambient temperature. This feature gives the rate of rise detector an advantage over thermal detectors because it is not affected by thermal lag.

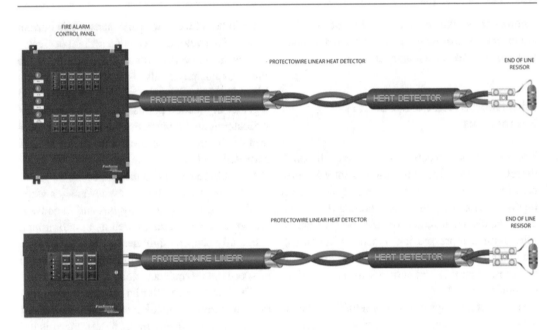

Fig. 3.2 Protectowire Linear Detector. (Courtesy of Protectowire)

When a thermal detector with a fixed operating temperature reaches that operating temperature, the air surrounding the detector is considerably higher than the detector-set temperature. This difference in surrounding air temperature and the operational temperature of the unit is due to the rate at which the temperature rises, and heat is lost in both the body of the unit itself and the surrounding area.

The rate of rise detector is activated by the very rate of temperature rise that is lost with a thermal detector. If the temperature rises at the rate of 15° per minute, or faster than 15° per minute, the rate of rise detector will activate.

Certain disadvantages of the rate of rise detector must be considered. Slow-burning or smoldering combustion may produce considerable heat, and the ceiling where the detector is located may rise to a very high temperature, but unless the rate of heat rise is rapid enough to produce a 15° per minute increase, the detector will not activate. It is for this reason that combination fixed temperature and rate of rise detectors are used to overcome the rate of temperature rise disadvantage. The combination fixed temperature and rate of rise detector are activated when the

rate of temperature rise is 15° or more per minute. It also has a fusible element with a predetermined operating temperature.

The combination electrical spot detector is a hollow shell containing a diaphragm. When the shell is subjected to combustion heat that is increasing in temperature at a rate of 15° per minute or greater, the air trapped in the detector shell expands, thereby forcing the diaphragm up to make an electrical contact that closes a current and activates the system. The system may be an alarm system, a suppression system, an equipment shutdown system, an equipment start-up system, or a combination of any of these systems. By closing the electrical circuit, the detector signals the control panel which transmits a signal, or signals, to the various systems.

There is a small vent on the detector to vent small pressure changes within the shell due to changes in the environment where the detector is located. These environmental changes may include starting up the area heating system, or sun on the metal roof after a cool night, or any number of situations that cause a slow rate of temperature increase. These increases are not in the 15° per minute range, but would cause a slow increase in pressure within the detector shell which could eventually depress the diaphragm and activate the detector if not vented.

The vent is calibrated so as not to be capable of venting sufficient pressure increase when the temperature rise of 15° per minute increases the pressure to the trip point.

Except for the fusible element that melts and is released from the detector at a fixed temperature, the detector in the rate of rise mode is self-restoring. When the fusible element reaches its operating temperature and is released from the detector, it releases a spring that mechanically depresses the diaphragm and activates the detector.

Ionization Versus Photoelectric Spot Detection

In evaluating the hazard, the level of required detection sensitivity, the environment in which the detector will function, and the type of combustion anticipated, it may be determined that the combustion should be detected in the incipient stage before there is appreciable heat and smoke in order to prevent heat and smoke damage to delicate electronic equipment. In such cases, ionization detectors should be seriously considered because they react when invisible products of combustion are given off by the combustion before there is visible flame, or heat development at the ceiling level, and little or no visible smoke has developed.

The analysis of the combustibles present and the type of fire that these combustibles will produce should be the first concern when making a detector selection. Fires that develop rapidly into the flaming stage without the accumulation of heavy, visible smoke particles are detected most effectively by ionization detectors. A photoelectric smoke detector is activated by visible smoke particles entering the detector. When a smoke detection system is proposed, it is recommended that a determination be made whether ionization detectors or photoelectric smoke detectors would be best suited to the hazard conditions. It is not just a matter of terminology but is important to differentiate between ionization and the photoelectric detector when recommending either type. The public generally refers to both the ionization and the photoelectric detector as smoke detectors, but when an engineer is recommending a detection system based on the results of an engineering evaluation, it is very important that the terminology be correct.

Ionization Detector Operation

The ionization detector contains a minute amount of radioactive material that ionizes the detector chamber. In other words, the radioactive material emits sufficient material to charge the atmosphere in the chamber of the detector to develop a current flow through the air space separating two electrodes. When minute invisible products of combustion (and/or small smoke particles) enter this electrically charged chamber, they attach themselves to the charged radioactive particles in the current flow, and by so doing, reduce the current flow between the two plates. This is shown in Fig. 3.3.

When the flow level between the plates is reduced by these heavy particles to a predetermined level, the minute voltage change is amplified and the detector is activated. This is a simplified explanation of how the ionization detector functions and is offered to further illustrate the difference between an ionization detector and a photoelectric smoke detector.

Photoelectric Detector Operation

A photoelectric smoke detector works on an entirely different principle than the ionization detector. One type of photoelectric smoke detector uses a light source that is projected onto a photosensitive device. When visible smoke particles enter the detector and pass between the light source and the photo-sensitive receiver device, they reduce the light reaching the receiver. This causes a slight voltage change, which is then amplified, and the detector is activated. Another type of photoelectric detector

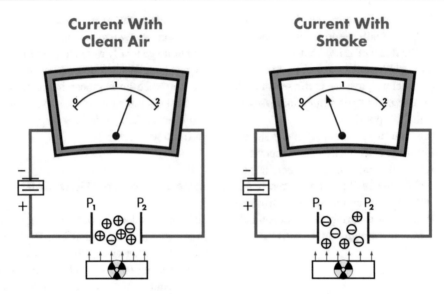

Fig. 3.3 Ionization Smoke Detector Operation. (Courtesy of John Morrison)

uses the light source and the photosensitive light receiver, but with this particular detector, the light source does not fall on the receiving sensor. When visible smoke particles enter the detector, the light is scattered by the smoke particles and in addition, the light is reflected from the smoke particles. When enough of this light reflected strikes the sensor receiver, the voltage change is amplified, and the detector activates. This is shown in Fig. 3.4. The photoelectric detector responds to the thicker, heavier, visible smoke that develops when slow-developing and smoldering combustion is present. A slow-burning or smoldering fire is typical of combustion of materials found in the office, home, hotels, dormitories, and hospitals, since this type of combustion produces the type of smoke created by this type of occupancy.

It is important to realize that the selection of the proper detector is an extremely difficult decision because of the many variables that sometimes make it almost impossible to determine with accuracy the nature of the potential fire. Every modem building contains equipment and furnishings manufactured of different, and sometimes unique, materials that may produce a type of combustion that is not anticipated in evaluating the proper detector for the area. Although the cost of a detection system will be increased, an engineer

may, in order to provide reliable combustion detection, recommend the installation of both ionization and photoelectric detectors if the type of fire potential cannot be readily determined.

In an effort to obtain a sensitive, early warning detection system where ionization detectors appear to be the best selection, it is imperative that the general environment be considered because these detectors may react to many non-fire invisible particles, including exhaust fumes, welding operations, kitchen odors, and pilot lights. Almost any detector can be activated by electrical equipment of a certain voltage and radiation level. Hospitals may have problems with detectors exposed to CAT scanners, high frequency radio transmissions, etc.

Maintenance is a prime consideration with detection systems. Dust and contamination can affect the sensitivity of any detector, either by making it more sensitive to non-fire detection sources, or by reducing the sensitivity.

Linear Beam Smoke Detector

The linear beam smoke detector basically operates on the same principle as the photoelectric smoke detector with a light source directed on a photo-sensitive receiver. The difference is that

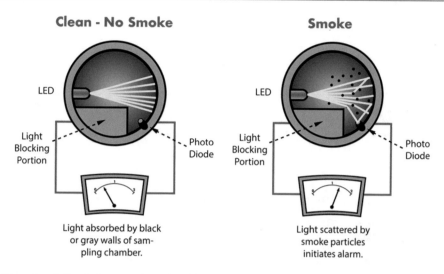

Clean - No Smoke

LED

Light Blocking Portion

Photo Diode

Light absorbed by black or gray walls of sampling chamber.

Smoke

LED

Light Blocking Portion

Photo Diode

Light scattered by smoke particles initiates alarm.

Fig. 3.4 Photoelectric Smoke Detector Operation. (Courtesy of John Morrison)

the linear beam smoke detector consists of two separate units. The light source is one piece of equipment, and the light beam receiver is a separate piece of equipment. The linear beam smoke detector light source, or transmitter, emits an invisible infrared beam over open area distances of 35′–300′ onto the light receiver.

When no smoke interferes with this beam, the receiver accepts the beam at a specified voltage level, but when smoke interferes with the beam, the infrared light reaching the receiver is lessened, and when the beam intensity drops below the predetermined sensitivity level of the receiver, it initiates a signal from the detector. This is shown in Fig. 3.5.

This type of smoke detector has many applications in areas where spot-type smoke detectors could not be installed to operate efficiently, such as atriums, airport terminals with very high ceilings, aircraft hangars, churches, and large open facilities. Spot-type smoke detectors installed on the extremely high ceilings of these areas would be almost inaccessible for maintenance, and since the beam transmitter and the sensor receiver are installed on the walls, the architectural beauty of a church ceiling or hotel atrium will not be compromised by the appearance of several spot detectors. One special application of a linear beam detector is in psychiatric facilities and cor-

rectional institutions where spot-type detectors would be subject to vandalism.

Rate Compensation Detector

The rate compensation detector, sometimes referred to as a rate anticipation detector, consists of two metallic struts mounted inside of a steel elongated shell. The stainless-steel shell has a coefficient of expansion greater than the two metallic struts mounted within it. When the shell is exposed to heat, it expands, and since the ends of the two struts are attached to either end of the shell, the expanding shell pulls the struts and stretches them until they make contact at the center of the struts. Rate compensation detectors have a predetermined fixed temperature set point, and when the two struts make contact and activate the detector, it has reached this temperature set point. Unlike the rate of rise detector that requires a minimum of 15° per minute rate of temperature rise to operate, a smoldering fire producing a slow rise in temperature will heat both the outer shell and the struts equally until the set point of the detector is reached. When the combustion heat rate is rapid, the rate compensation shell expands rapidly until the struts are stretched to the point of contact, thus eliminating the thermal lag that may prevent a

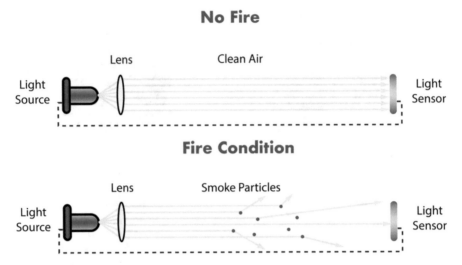

Fig. 3.5 Linear Beam Smoke Detector Operation. (Courtesy of John Morrison)

fixed temperature detector from operating when the surrounding air has reached the set point of the detector.

Duct Detectors

Duct detectors are either ionization or photoelectric. They are usually installed on a return air duct close to the fan unit. Duct detectors are installed on the exterior of the duct with long tubes extending into the duct. These tubes have holes that pick up products of combustion in the air stream returning from the vents in the facility, transmitting them into the detector unit on the outside of the duct. (See Fig. 3.6.)

When the detector is activated, one of the signals it actuates is an automatic shut-down of the heating and air-conditioning fans to prevent the products of combustion from being sent back through the heating and air-conditioning system throughout the building.

Duct detectors provide an excellent function, but always keep in mind that they should not provide a substitute for a detection system throughout the facility. The return air to the fan units may come from many parts of the building, and smoke generated in one room of the building and sucked into the return air vent of that room may be so diluted by return air from other rooms where no

combustion is occurring that by the time it reaches the duct detector, it is not concentrated enough to activate the detector.

Flame Detectors

Flame detectors respond to the visual radiant energy of a flame. A flame produces visible wavelengths of radiant energy, and different flame detectors see different radiant wavelengths given off by a flame. There are two basic types of flame detectors. One is ultraviolet, and the other is infrared. The ultraviolet flame detector responds to ultraviolet flame wavelengths, and the infrared responds to infrared flame wavelengths (Fig. 3.7).

Ultraviolet flame wavelengths are the shorter wavelengths emitted by high intensity flames, but this range of wavelengths is also emitted by the sun, so when using an ultraviolet detector, it must be designed to be "solar blind".

The ultraviolet detector sensitivity decreases with the distance from the flame source, whereas the infrared detector will function effectively when separated some distance from the flame source. This is a generic statement, backed up by the actual fact that ultraviolet detectors installed in an aircraft hangar have been activated by welding operation several thousand feet from the hangar.

Fig. 3.6 Duct Type Detector. (Photograph by Robert Till)

Fig. 3.7 UV Detection in Subway Booth. (Photograph by Robert Till)

Flame detectors are quite specialized detectors and a considerable level of expertise is required to design a system that is free from activation by non-fire sources, yet sensitive enough to activate at the first flicker of a flame. These detectors are also extremely sensitive, and a system using them must be designed with great care by experienced engineers. A rotating beam light on top of a truck, a slowly rotating fan blade, or the heat from an aircraft engine can activate a flame detector.

In order to attempt to reduce false activation, the ultraviolet and infrared detector can be used in tandem. Both must be activated in order to activate alarms or trip systems. It is anticipated that a non-fire ultraviolet source will not occur at the same time as a non-fire infrared source, but that a flame in a flammable liquid will produce a flame front that will emit both ultraviolet and infrared wavelengths sufficient to activate the detector.

The ultraviolet detector has a viewing area resembling a cone. This cone of vision is approximately 90° closer to the detector and widens out to approximately 60° at some distance from the detector. The detector sensitivity decreases as the distance from the flame source to the detector increases. For example:

A one-square-foot hydrocarbon pan fire can be detected when located some 50 from the detector, and a nine-square-foot hydrocarbon fire can be detected some 100 from the detector.

As the fire size increases, the detector can be activated with the flame front at greater distances from the detector. It must, however, be kept in mind that the sensitivity of the detector is decreased as the flame front moves away from the center line of the detector. This is particularly important when designing floor coverage detection with the ultraviolet detector. The cones of vision of the individual detectors should overlap so that a flame anywhere in the area will be within the sensitive area of the detector cone of vision.

The infrared detector also has a cone of vision similar to, but not as wide as, the ultraviolet detector. As with the UV detector, the IR detector is most sensitive to flame detection when the flame source is directly in line with the detector. The IR detector has only has an 80% sensitivity when the flame source is 30° off the center of the detector, and 50% sensitivity is accomplished when the flame source is 45°, off the center line of the detector (Fig. 3.8).

When flame detectors (optical detectors) are selected for the detection system, and the system is to activate suppression systems, it is recom-

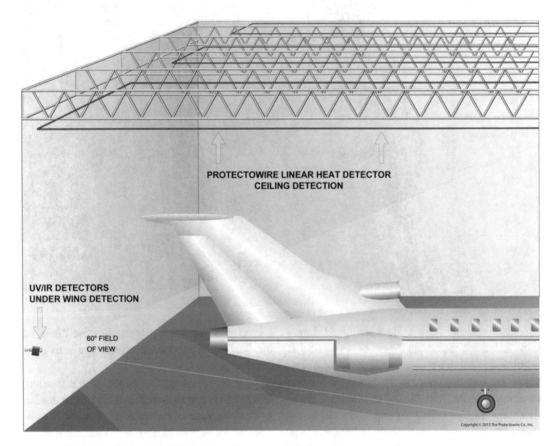

Fig. 3.8 UV/IR and Protectowire Application. (Courtesy of Protectowire)

mended that one detector activation operate the alarm system, and the second detector operate the suppression system.

Flame, or optical, detectors are extremely sensitive to flaming fires, but are also very sensitive to non-fire sources of activation. When a flame detection system not only activates an alarm system, but also automatically operates a suppression system, it may be advisable to monitor the optical system for a short period of time after installation.

This monitoring is best explained using the example of aircraft hangar protection: When the foam systems in a large aircraft hangar are activated, they will discharge a large amount of foam over a wide area. This expended foam is expensive to replace, and the labor involved in replacement can also be time consuming and costly. The cost of foam replacement is actually minor compared to the cost impact of discharged foam removal from machinery, electrical equipment, HVAC units, and the aircraft parked in the hangar.

The sensitivity of the flame detectors is critical to the protection of the aircraft undergoing maintenance in the hangar, and to the hangar structure itself. A jet fuel spill fire under the aircraft must be controlled by the suppression system within 30 s and extinguished within 60 s. Otherwise, the aircraft could be damaged beyond repair.

An ultraviolet or an infrared detector can look under the wings of an aircraft with the sensitivity to trip the suppression systems within the required time frame. But it is this very sensitivity to non-fire sources of activation that can cause the serious and expensive problem of false dumps of the suppression systems.

Once the installation of the detection and suppression systems is complete, the detection system is connected to a monitor, having the signals from the detection system interlocked with the suppression systems.

When the detection system is activated, the alarm system becomes operational, and if a fire is observed, the suppression systems can be activated by operation of a manual pull station. The detection system also activates the monitor which prints out a date and time of the activation. The non-fire source of activation can usually be determined. Having a record of the causes of false detector activations will allow corrective measures to be instituted before the suppression systems are placed on automatic activation. Monitoring has an initial cost, and the manual activation of the suppression systems in a fire situation requires prompt attention. Nevertheless, if the monitoring keeps an area as free as possible from non-fire-related sources of detector activation, the initial cost of the monitoring is negligible compared to that of an unnecessary system operation.

The latest infrared technology of flame detectors (IRRR) provides a very high level of sensitivity in detecting flames from hydrocarbon-based liquids and gases within a wide field of view. The technology ignores false triggers from sources where UV in other detectors may be a problem. This includes areas such as direct or indirect sunlight, welding, resistive heaters, fluorescent, halogen, and incandescent light.

Air Aspirating Smoke Detection Systems

Air aspirating smoke detection systems, often called VESDA systems (after one of the first companies to produce them) provide very early warning of smoke conditions. Suction tubes are used to draw air for different locations within a protected space. Air drawn into a smoke detector or gas-analysis device determines if fire is present (see Fig. 3.9). These systems are very sensitive, and often tubes are mounted inside electrical cabinets and other spaces where rapid detection is necessary to prevent damage. They are often used in conjunction with halon alternatives to keep the fire size and therefore the amount of hydrofluoric acid to a minimum in situations where these suppression systems are used.

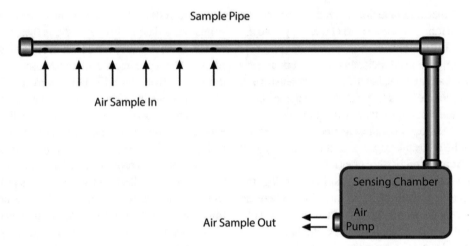

Fig. 3.9 VESDA type system operation. (Courtesy of John Morrison)

Notification

4

History of Notification

The Telegraph

Fire alarm systems that report either on the premises being served or to parties remote from the original premise are direct descendants of the first viable commercial telegraph in developed in 1837 by Samuel Morse. In 1844 the first commercial telegraph line was constructed between Washington DC and Baltimore. The telegraph was the "internet" of its day, allowing text to travel from city to city on wires at the speed of light. Limitations were only the speed at which they could be coded and decoded.

Fire Alarm

In 1847, the New York City commissioned the first fire alarm system based on the telegraph. The system, required by ordinance "to construct a line of telegraph, by setting posts in the ground ... for communicating alarms from the City Hall to different fire stations and to instruct the different bell-ringers in the use of said invention" (Greer 1991).

Cornelius Anderson, the New York Chief Engineer, distilled the hopes for the new system in his annual report of 1947:

From the Proceedings of the Board of Assistant Aldermen: Volume 30:

> ...This system would ...[prevent] nearly all false alarms, and at the same time transmit the alarm in case of fire to all parts of the City with such certainty and promptitude that it would be the means of saving annually to the city a sum nearly equal to the cost of its erection... and maintenance; it would also very greatly diminish the labors of the Firemen, for in the event of a fire, the policeman on whose post it occurred would give the alarm immediately to the station house, and it would then be transmitted to the nearest bell station by telegraph. (Ditzel 1990)

However, the first system developed in New York City had a number of shortfalls:

- All 8 towers, including the city hall watch tower, as well as the belfries with bells, were linked together in a single telegraph circuit, and therefore the system lacked a central station.
- Anyone who detected a fire needed access to the alarm telegraph as well as the wherewithal to use it -not a very common thing.
- The system could not distinguish between two separate alarms, a common occurrence in a city like New York.

An improved telegraphic fire alarm, suitable for development at the commercial scale, was first developed when Channing and Farmer constructed a system in Boston in 1852. This system consisted of alarm boxes located in various districts that all reported to the same central station within the city. Fire companies could be dispatched based on the

© Springer International Publishing AG, part of Springer Nature 2019
R. C. Till, J. W. Coon, *Fire Protection*, https://doi.org/10.1007/978-3-319-90844-1_4

activation of particular boxes within the district. The fire departments could report to the particular box and determine where the nearby fire was from there.

The Influence of John Gamewell

John Gamewell became aware of the system developed by Channing and Farmer. He bought the rights to the system first for the south and west, and later for the entire US. John N Gamewell and Co. was formed in 1859. Gamewell quickly took control of the newly formed fire alarm industry by purchasing patents that covered most of the key components. Up until 1861, Gamewell managed to sell systems to Philadelphia, St. Louis, Baltimore, New Orleans and Charleston S.C.

The American Civil War, from 1861 to 1865 nearly ruined Gamewell. As a Southerner he lost all of his patents for fire alarm systems. He moved to Hackensack, New Jersey, after realizing that moving his business forward in the south would be difficult. John Kennard of Boston purchased Gamewell's patents after the war and returned most of them to him. The firm of Gamewell, Kennard and Co. was formed in New York in 1857.

In 1871, Gamewell patented a design that eliminated interference when two boxes were pulled at the same time. This occurred when two boxes were pulled on the same fire by two different individuals. The growth of central station fire alarm systems is summarized in the Table 4.1.

After 1904, the Census Bureau reported there were 764 systems operating in the US.

Gamewell effectively perfected the central station and auxiliary alarm systems. Systems were activated by fire alarm boxes, heat detectors or sprinkler systems.

Telephone Systems

In 1876, Alexander Graham Bell developed the practical telephone. It was able to transmit fire alarm signals and notify firefighters. It had the

Table 4.1 Number of systems installed per decade

Years	Number of systems
1852–1962	4
1862–1872	40
1872–1982	60
1882–1892	299
1892–1902	359

drawback that most families that adopted the phone had party lines (multiple phones on the same line). This could develop some confusion as to where the fire actually was. It should be noted that the first equivalent of 911 service was developed in 1937 in the United Kingdom. The development of the telephone simplified other variations on how a fire might be transmitted. These include the remote alarm system and the proprietary alarm system and provided more confidence in the local alarm system.

911 System

The US did not have a universal phone number before the 1960's. In the event of an emergency (fire, police etc.) callers had to know the phone number for each police department or fire department that needed to respond. In the days before automatic switching, the process may have been held up by an operator who was busy with another call. (Stone 2014)

In the case of large cities this situation was exacerbated as there might be 50 different police or fire departments and just as many numbers. Telephone operators were sometimes left to direct calls if the caller wasn't sure which firehouse or police department was theirs.

In 1957, to deal with these issues, a national emergency number was suggested by the National Fire Chief's Association. A decade later a report to President Lyndon Johnson's Commission on Law Enforcement and Administration also recommended a single national number for citizens to call in the event of an emergency.

The Federal Communications Commission (FCC) partnered with the American Telephone and Telegraph Company (AT&T) to determine what number should be used. AT&T proposed the numbers 911. The technical reasons that these numbers were chosen are simple – they are short, easy to remember, and can be dialed quickly on the rotary phones of the day. In addition, AT&T had started to use 611 and 411 numbers for reporting telephone service problems and for providing telephone number information respectively, so a new number of this type could be easily accommodated by the existing phone switching hardware.

About a quarter of the US had 911 service about 10 years after congress established the number as the universal emergency phone number. In 1989 that number rose to about half. Today almost all people in the US have 911 service.

Internet and Addressable Systems

The internet began as a reliable military network that could tolerate losing nodes of communication. It spawned another revolution in the civilian sector – the modern addressable fire alarm system. The method of addressing individual devices in on the network ultimately became a network layer that detection devices could converse on. At first this prospect was extremely expensive, but as time and economies of scale improved the prices of these devices, it is now cheaper to use addressable devices in a fire alarm network than it is to use a zoned approach.

Notification System

Fire Alarm Systems

It is extremely important that detection systems that alarm and operate suppression systems be designed with every effort to prevent false activation of the alarm (notification) and/or the suppression systems, without decreasing detection sensitivity below the point at which it reads effectively in a fire situation. Systems include a fire alarm control unit, primary and secondary power supplies, and various circuits used to connect to detection (input) devices, notification (output) devices or both (known as signaling line circuits).

Fire Alarm Control Unit

A fire alarm control unit is defined by NFPA 72 (2016) as "A component of the fire alarm system, provided with primary and secondary power sources, which receives signals from initiating devices or other fire alarm control units, and processes these signals to determine part or all of the required fire alarm system output function(s)".

Conventional FACU

A conventional Fire Alarm Control Unit (FACU) has one or more circuits connected to initiating devices, which are wired in parallel. The initiating device circuits are zoned so that fires can be located within the building. An initiating device circuit will show a zone in one of three states; normal, trouble, or alarm.

Addressable FACU

Modern addressable systems are essentially large computer network with each device having its own address. They began to appear on the market during the 1980's during the initial microprocessor boom. When a detector goes into alarm, the fire alarm control panel is able to locate exactly where the fire is by essentially looking up the devices address on a table and can direct the fire department directly to the location of the fire. Advantages of addressable systems include:

1. Upon arrival at a fire, firefighters can quickly identify the specific detection devices that have activated.
2. The devices and appliances can continuously inform the panel that they are working correctly. In other words, the panel can supervise them.
3. The exact location of every activated device can be determined by consulting the FACU because each device has its own address.
4. Devices or appliances need to be serviced are easily located, because they can report their own failures to the FACU, or simply will not "talk" to the FACU.

5. If system designers would like certain appliances such as horns or strobes in specific areas to activate under certain conditions this is easily achieved because only certain network addresses can be programmed to activate under these conditions.

Analog Addressable System

Analog addressable systems are the latest system available as of the publication of this book. They operate in analog mode in terms of how they detect fires. For example, they can provide precise measurements within a protected area. They may report that the temperature in some area of a building is 70 °F while in another area it is 72 °F. These systems allow for increased flexibility in setting an alarm condition. For example, a system may activate in a "pre-alarm" mode when the temperature reaches 120° at a specific detector and then go into full alarm mode when the temperature reaches 135°.

These same units operate digitally in terms of how they are connected together. They each have their own "address" so that the precise location of the detector can be known.

Supervision

Supervision of detection and alarm systems is a key word in the design of all systems. The word control used in the following description of system supervision refers to all of the action that a signal from a detection system may control or activate: suppression system activation, equipment shut-down, equipment start-up, magnetic door closers, etc.

Detection, alarm, and control systems that are electrically operated must have wiring supervised to activate a trouble alarm should a wire break, short out, become unintentionally grounded, or should a detector malfunction. Usually, the trouble signal activates a light and an audible alarm at the control panel or annunciator, and a silence switch shuts off the audible trouble alarm, but the light will remain on until the trouble is corrected.

Most equipment reactivates the audible alarm when the trouble is corrected, and the audible alarm will not be silenced until the silence switch is returned to active position, indicating that the panel or annunciator has been returned to operational condition. Supervision for detection, alarm, and control systems is imperative, because a system that has been rendered inoperable without any indication of its condition, will provide a very serious false sense of protection. Pneumatic detection, alarm, and control systems must have a low air alarm device to activate a trouble alarm when the air pressure drops below operational pressure.

Other Issues

Another system installation feature that must be considered in evaluating the area where the system is to be installed is the possibility of flammable vapors or dust conditions, both of which could produce an explosion when the atmosphere is exposed to an electrical spark. When this condition is encountered, the detectors, wiring, and equipment must be specified as explosion-proof.

Explosion-proof equipment and wiring is designed to prevent any spark-producing connections from being exposed to the atmosphere where it could initiate an explosion. Explosion-proof equipment and wiring are more expensive to purchase and install; therefore, it is important to always make this determination when specifying and estimating a detection, alarm, and control system.

Another feature of detection, alarm, and control systems that is extremely important to include in the design, specification, and estimate sheet, is battery back-up. If the system does not have a separate, reliable, and uninterruptible source of power to use when the primary, commercial source of power is lost, the system must have a battery back-up power source. Battery back-up to operate the system and all of its components during loss of primary power must be arranged and equipped to come on line automatically, and within a matter of seconds, when the primary source of power is lost. The batteries should be kept fully charged by an automatic battery charger, and the battery power should be sufficient to operate the entire system for an extended time (which is usually 24–48 h). The control panel or annunciator must have an indicator light which shows that the system is operating on battery back-up, and should also have an indication that the batteries are in a charged condition. On the subject of power supplies, supervision, and wiring of detection, alarm, and control systems, the National Electric Code (NFPA Standard 70) requirements must be followed.

Types of Fire Alarm Systems

There are a number of different ways to notify the inhabitants of a building and the fire department, so they can each respond to a fire. The most obvious for the fire service is via 911 service as discussed. Other standard systems are described here.

The components used to detect the fire have been described previously. They include smoke and heat detectors, flow switches from sprinkler systems, and other means.

Local Alarm System

A local alarm system produces alarms only for the facility it serves, or the premises of the area served. This type of alarm system may be used in schools, hospitals, institutions, and industrial facilities where the alarm is audible, and sometimes also visible, within the building or facility it protects. In this case notification of the fire department is a human operation by telephone or other human means.

Auxiliary Alarm System

The auxiliary alarm system activates local alarms, but this system is also connected to a municipal fire alarm box so that when the alarm system is activated, the municipal fire alarm box is also activated, which in turn transmits a coded signal directly to fire alarm headquarters. This coded signal received by the fire alarm headquarters identifies the individual fire alarm box transmitting the signal. The fire department can be directed to the specific location of the fire alarm box.

Remote Alarm

This is a very effective fire alarm system which, when activated, automatically transmits a signal to a constantly attended location. A constantly attended location is most often the most important facet of any fire alarm system, as is the automatic transmission of this remote alarm signal. Any time the human element is introduced, there can be delays in summoning the fire department which can spell the difference between successful lifesaving operations and extinguishment of the combustion, or loss of life and total destruction of the property.

One relatively simple and inexpensive method of transmitting a remote signal to a remote alarm device is over a leased and dedicated telephone line. The remote alarm device can be located in a fire station, or any facility where someone is in constant attendance. The control panel should be specified as having contacts for the automatic transmission of remote alarms, and the transmission of trouble alarms, which can also be incorporated into the panel. Usually when the remote signal is transmitted to a municipal fire alarm headquarters or fire station, the authority having jurisdiction must give permission to transmit trouble signals.

Usually a municipal fire alarm remote device will receive only alarm signals, since the receipt of the signal and correction of trouble on the system should be handled by the owner of the system. If life safety is a major factor, the municipal fire alarm department may allow trouble signals to be automatically transmitted as well. If the remote signal is transmitted to a municipal station, a requirement is imposed to have the facility owner sign a contract to have the systems remote alarm tested on a monthly or periodic schedule. This requirement is intended to prevent false alarms, with the resulting responses from the fire department. This additional ongoing cost should be realized by the client or owner of the detection and alarm system.

Proprietary Alarm System

The proprietary alarm system serves a facility or complex of facilities with its own alarm-receiving location. An example is a complex of industrial facilities equipped with detection and suppression systems, where every system activates local alarms and also transmits an alarm to a central, constantly attended area that is a part of, and owned by, the complex. This system is an independent alarm and alarm notification system under one ownership. The alarm-receiving station also receives trouble signals, fire pump signals, etc. Usually the alarm and trouble signals are received by an audible and visual indicator. The signal location is typically indicated on a graphic annunciator, or an annunciator that indicates which system or equipment has transmitted the signal. The response can then be immediate to the area where the signal originated.

Central Station

A central station system is a commercial system where alarm signals are automatically transmitted to a central station that employs experienced personnel whose only occupation is the monitoring of the panels that receive the signals. A central station usually has many clients, and when an alarm or trouble signal is received, the panel indication will inform the operator exactly which client facility is transmitting the automatic signal. Central station firms receive security and other monitoring alarms, and the operators are trained to interpret the signals, dispatch their personnel to investigate the alarm, and also to notify the fire department and/or police department. Naturally, there is a monthly charge for this central station service, and sometimes these central station companies require that their own equipment be installed to transmit the alarms in order to provide adequate integrity of the equipment and to avoid the possibility of inadequate equipment transmitting numerous false alarms.

References

Ditzel, Paul. *Fire Alarm!: The Fascinating Story Behind the Red Box on the Corner (Fire Service History Series)*. Squire Boone Village, 1990.

Greer, William. *A History of Alarm Security*. Bethesda, MD: National Burglar and Fire Alarm Association, 1991.

Stone, Sarah. "How 911 Became the Emergency Call Number." 12/4/16 <https://gizmodo.com/how-911-become-the-emergency-call-number-1601064956> (2014).

Fire Pumps and Water Supplies

5

Early Methods of Suppression of Fires and Conflagrations

The first recorded attempts at fire suppression involved the use of water as an extinguishing agent to control an uncontrolled fire. Leather buckets filled with water were kept on hand by the householder to extinguish fires in the home, or to wet down combustible roofs to prevent the spread of fire to adjacent buildings from exposure to sparks and radiant heat.

In 1723, Ambrose Godfrey devised an improvement over merely throwing a bucket of water or sand on a fire. Godfrey's arrangement consisted of a barrel filled with water which also held a charge of gunpowder in a waterproof container. The idea was to light a fuse and then throw the barrel into the fire. The fuse ignited the gunpowder and the explosion shattered the barrel, scattering the water on the fire. Crude as it may seem, the Godfrey Barrel was the first attempt to convey extinguishing water to the seat of the fire-an area that previously could not be reached because of intense heat and flames.

The next step in the evolution of fire suppression systems was the automation of water delivery through a piping system, as opposed to applying the extinguishing agent to the fire by hand.

Fire Pumps

Fire pumps are an integral part of fire protection, and as such, must be part of the working knowledge of anyone designing, specifying, or estimating the cast of fire suppression systems. This section will address the two most common fire pumps: the diesel drive fire pump and the electric drive fire pump.

When the municipal water pressure available is not sufficient in pressure or volume for a sustained fire extinguishing operation, then other sources must be found. If the water is available, but not at sufficient pressure, then fire pumps are used to maintain the required pressure and volume. In situations where there is no reliable water source, the use of storage tanks and reservoirs are warranted.

If the pressure demand of the suppression systems exceeds the city mains capacity, the first consideration is the use of a fire pump to boost the pressure. If, on further examination, it may be found that the booster pump can furnish the required demand pressure, but at the volume (gpm) demand of the suppression systems, the city main pressure will be below the 20 psi residual pressure necessary to allow the fire department pumper trucks to utilize the hydrants. On the other hand, the supply main may not be large enough to furnish the volume demand of the suppression systems. In either case, the use of a

© Springer International Publishing AG, part of Springer Nature 2019
R. C. Till, J. W. Coon, *Fire Protection*, https://doi.org/10.1007/978-3-319-90844-1_5

stored water source to supply the fire pump and system demands is warranted. For example:

> A large company builds a large manufacturing and rack storage warehouse in a small town. Because of the extensive rack storage system, the demand for the suppression systems far exceeds the supply offered by the city water system. The result is a storage tank dedicated to supplying the fire protection filled from the city water system and furnished to the suppression systems through the fire pumps. Actually, two storage tanks of an equal volume are required, each dedicated to fire protection supply. This requirement is based on the concern that if one tank were out of service for repair and painting, the facility would be without fire protection. This is a very important consideration in recommending and designing reliable fire protection water supplies.

Elevating the storage tank is one way to attain the required pressure. Another is with the use of fire pumps. This chapter provides a basic understanding of fire pump design, devices, and equipment, and the difference between fire pumps and mechanical pumps. It also provides an overview of tanks and their use in fire protection.

Each pump manufacturer submits their pump to a nationally known testing laboratory such as Underwriters Laboratories (UL) or Factory Mutual Research (FM). These laboratories run exhaustive tests, and if the pump and all the accessories meet their standards, the pump obtains the UL listing label and/or the FM approval. The requirements for such listings are costly, and therefore, a listed or approved fire pump is considerably more expensive than a hot water or chilled water circulating mechanical pump of equal size. Estimators, designers, and specifiers should always be certain that water supply pumps are UL listed and/or FM approved.

It is often said among mechanical engineers that the only difference between a fire pump and a mechanical pump is that the fire pump is painted red. Actually, it takes a great deal more than red paint to create a fire pump. Not only is the cost higher (the pump manufacturer invests a considerable amount in obtaining the Underwriters Laboratories and/or Factory Mutual Research tests necessary to obtain the listing or approval), but the fire pump performs with totally different criteria.

The chapter Regulatory Agencies, Authorities and Organizations discusses the requirements that insurance carriers place on the use of UL listed and/or FM approved fire pumps to meet the water supply volume and pressure required to maintain operable and safe fire protection systems. Insurance carrier requirements are based on how the pump is constructed, successful test results, and the performance of the pump in discharge volume at various pressures. The carrier has confidence in a pump that, except for periodic testing, may lay dormant for years, and then have to operate at 100% efficiency when activated in a fire condition. The main distinction between laboratory testing requirements lies in the type of motor. When a fire pump is to be driven by a diesel engine, both the pump and the engine must be UL listed and/or FM approved. When a fire pump is to be driven by an electric motor, only the pump must be UL listed and/or FM approved.

General Criteria

First, we will define a few terms that identify the types of fire pumps.

Centrifugal Pump: The discharge pressure is developed by the rotation, or centrifugal force, of the water.

Horizontal Split Case Centrifugal Pump: A fire pump that has the suction and discharge in a horizontal plane, and is denoted by the pump casing, or housing, split horizontally.

Vertical Shaft Turbine Pump: This is also a centrifugal type pump, but the pump is a vertical column connecting the impellers that drive the water up the column to be discharged from the top, or head, of the pump column.

Basic Design Feature of Fire Pumps: A fire pump will deliver 150% of the discharge volume (gpm) it is designed to deliver, at 65% of the design pressure (psi). For example:

> A fire pump designed to deliver 2,000 gpm at 100 psi. At 3,000 gpm discharge, this pump will be discharging at 65 psi. The design gpm and psi, and the 150% gpm at 65% psi, are the two fixed points on the pump curve. There is one other fixed point on the pump curves-shutoff pressure-that reflects the gpm and psi with the pump operating against a closed valve. Shut-off pressure is sometimes

referred to as the churn pressure because the pump is discharging against a closed valve and is just churning. Shut-off pressure cannot exceed 140% of the pump design pressure.

Horizontal Fire Pumps

"Standard" approved fire pumps have rated capacities of 500, 750, 1000, 1500, 2000, 2500, 3000, and 5000 gpm, and pressure ratings from 40 psi to 200 psi. Standard is in quotation marks because this chapter will only be dealing with these common pump capacities although smaller capacity fire pumps are available for special uses. Horizontal fire pumps must have a suction supply providing water to the suction of the pump under pressure. This positive pressure on the supply to the pump suction must be available when the pump is operating at 150% of its design discharge, which is called operating at overload.

This is an important consideration when the pump is taking its supply from a city water main. It must be carefully determined that when the pump is operating at overload, or in other words is discharging 50% more water than its design capacity, it cannot be allowed to pull the pressure of the city main below 20 psi. In addition to the volume of water being discharged by the pump at overload, the anticipated flow from hydrants on the city water main must also be added. The combined pump discharge volume plus the anticipated hydrant volume must not reduce the pressure in the city main below 20 psi.

The basic reason for this requirement is that the fire department pumper trucks require the 20 psi minimum pressure to operate their fire truck pumps when they take suction from the hydrant. In addition, this is a margin of safety to prevent sucking the main dry which could result in collapse of the main or leaking of the fittings on the main. Fig. 5.1 shows the following components:.

1. Pressure sensing lines installed per code requirements
2. UL Listed/FM Approved diesel engine
3. Flowmeter loop
4. Listed OS&Y valve
5. Listed butter y valve
6. Listed main relief valve
7. Single point electrical connection
8. Pre-piped fuel system
9. UL Listed FM approved fire pump controller
10. Closed waste cone
11. Listed check valve
12. Stuffing box and engine raw water drain
13. UL Listed fuel tank sized per NFPA 20
14. UL Listed FM approved fire pump

Suction and Discharge

The size of the suction and discharge pipe, fittings, and valves is mandated by NFPA Standard 20, and by the Factory Mutual Data Sheets. (Factory Mutual Data Sheets are equivalent to the NFPA Fire Codes.) The following suction and discharge pipe sizes are minimums:

500 gpm fire pump 5″ suction and discharge
750 gpm fire pump 6″ suction and discharge
1000 gpm fire pump 8″ suction, 6″ discharge
1500 gpm fire pump 8″ suction and discharge
2000 gpm fire pump 10″ suction and discharge
2500 gpm fire pump 10″ suction and discharge
3000 gpm fire pump 12″ suction and discharge

Note: The use of fire pumps with capacities of 500 gpm through 3000 gpm is an example only. There are many fire pumps with various capacities up to and including 5000 gpm, and so-called special service pumps with capacities starting at 25 gpm.

Pump Suction

The control valve in the pump suction line must be an outside screw and yoke gate (OS&Y) valve. This type of valve will provide full and unobstructed flow into the suction side of the pump with little turbulence, where a butterfly type valve might not have the vane fully open, causing an uneven flow into the eye of the impeller. It is also recommended that the OS&Y valve be installed

Fig. 5.1 Diesel fire
pump unit on skid.
(Courtesy of AC Fire
Pump)

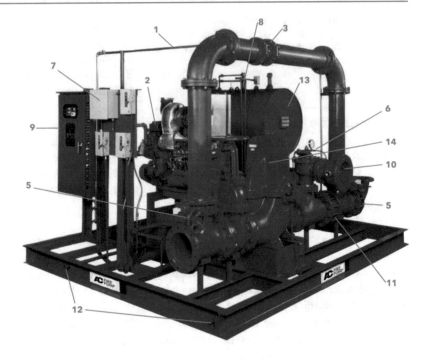

in the upright position to further guarantee the
even flow through the valve.

The one gallon per horsepower of the engine
for the fuel supply is based on one pint per horsepower per hour for 8 h. The fuel supply volume
should be verified by the engine manufacturer
before supplying the fuel tank.

Suction and Discharge Piping

It is good engineering practice to flange all of the
pump suction and discharge fittings and valves
for tightness and ease in replacement or repair,
although screwed and mechanical grooved joints
are acceptable. The suction pipe should be galvanized to prevent tuberculation. In lieu of galvanized suction pipe, the pipe may be painted
inside with an approved water-resistant paint.
All discharge piping is hydrostatical tested at
200 psi for 2 h.

Another important feature of the suction into
the pump is the fact that the suction pipe size may
not be the size of the flange on the pump suction.
In order to reduce the suction pipe size to bolt to
the pump suction flange, a flange decreasor must

be used. It is imperative that the reducer be an
eccentric rather than a concentric, to prevent any
possibility of air being sucked into the pump and
causing reduced suction and possibly creating
cavitation. Cavitation is a complex phenomenon
that can cause serious damage to the pump. The
use of the eccentric reducer minimizes the introduction of air into the pump at this point of connection. (See Fig. 5.2).

The use of an elbow with the center line parallel to the horizontal pump and installed at the
suction flange of the pump (or bolted to the
eccentric reducer) can create turbulence in the
pump and must be avoided. When multiple fire
pumps are installed, they usually take suction
from a large pipe header that supplies all the
pumps. If the pump suction pipe is smaller than
this header pipe, it is not good practice to connect
the smaller pump suction pipe to the center line
of the larger pipe header. The large pipe header
may not always be completely full of water, and
if the surface of the water in this pipe drops below
the top of the pump suction pipe, air will be
drawn into the pump. If this condition could
exist, then it is better to connect the pump suction
line to the bottom of the large pipe header.

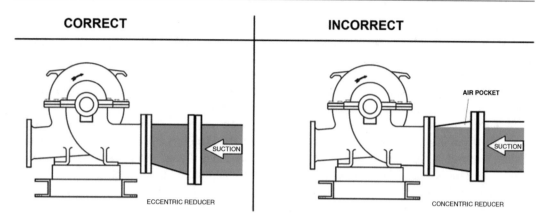

CORRECT   **INCORRECT**

SUCTION   ECCENTRIC REDUCER   AIR POCKET   SUCTION   CONCENTRIC REDUCER

Fig. 5.2 Eccentric pump reducer. (Courtesy of the AC Fire Pump)

All of these measures-including the eccentric tapered reducer and the OS&Y valve-are designed to maintain a constant flow of water to the pump, with a minimum of turbulence, since the introduction of air and/or turbulence in the flow can reduce pump efficiency.

Circulation Relief Valve

Every pump driven by an electric motor and radiator-cooled diesel motor must have a circulation relief valve, an automatic valve that provides sufficient water circulation when the pump is operating with no flow-against a closed discharge valve. If sufficient water is not circulated in this situation, the pump would keep churning the same water in the pump casing and the water would eventually overheat and damage the pump.

In designing the pump house or pump room, a drain must be provided to carry off the water discharged from the circulation relief valve. Fire pumps driven by a diesel motor without a heat exchanger cooling system do not require a circulation relief valve because the diesel engines water cooling system takes suction from the discharge of the pump.

When the pump is running against a closed valve, the diesel engine is obviously also running and taking water from the discharge of the pump and circulating it through the diesel engine to maintain the engine temperature. This diesel engine cooling water passes through a heat exchanger on the engine and is discharged. (This discharge must also be provided with a floor drain.) The circulation relief valve is relatively small – 3/4″ or 1″ depending on the size of the pump but the diesel engine cooling water could be discharging a considerable amount of water. It is best to determine from the diesel engine manufacturer how much water is anticipated before designing the drain size.

Main Relief Valve

Unless the system pressure is expected to exceed the design pressure (usually 175 psi), a relief valve is not required on fire pump systems where the pump is driven by an electric motor. System pressure could escalate if the suction pressure increased and subsequently increased the discharge pressure.

The electric motor is considered to be a constant drive motor, but the diesel engine is an adjustable speed driver, and also has the potential (although remote) of running wild and accelerating the pumps discharge pressure. A relief valve in the discharge line is sometimes required on all diesel drive fire pump systems, depending on the design.

The main relief valve is sized according to the size of the fire pump – a 3″ valve for a 500 gpm pump up to an 8″ for a 3000 gpm pump. To make sure there is no back pressure in the discharge pipe from the main relief valve, the relief valve discharge line is also quite large – 5″ for a 500 gpm pump up to a 12″ for a 3000 gpm pump. Although the main relief valve does not discharge more than a trickle of water when the pump starts, malfunction or excess pressure could expel hundreds of gallons per minute. For this reason, the most desirable location for this main relief valve discharge is outside the pump house or pump room onto grade. Discharging onto the floor of the pump house or room and into a floor drain can invite damage from splashing and the accumulation of water.

If the water supply to the pumps comes from a tank where the quantity of water is limited, the main relief valve discharge can be discharged back to the tank. Regardless of how the main relief valve discharge water is handled, the flow of water from the relief valve must be visible. This is accomplished by the use of a relief valve cone connected between the relief valve and the relief valve discharge line. This cone can be open or closed, with small windows to view the flow.

Pump Test

The fire pump must be tested by flowing the pump discharge after installation, and the insurance carrier may require an annual full flow discharge test. This flow test is accomplished in two ways.

1. The pump test line is connected to the pump discharge line after the discharge line check valve and ahead of the discharge line control valve.
2. The pump test line has a normally closed control valve and is piped to a test header installed on the exterior wall of the pump house or pump room. (See Table 5.1).

Pump Test Design

When the pump is to be tested, the discharge line control valve is closed, and the test line control valve opened, and the pump starts up. As each of the 2–1/2″ valves open, the flow is obtained and the pump discharge pressure recorded. Subsequent valves are opened until the pump is operating at design gpm and pressure, and then on up to overload – 150% of discharge and 65% of pressure.

There is another method of testing a pump. The pump test line is located as described above (in the pump discharge line), and the test line has a normally closed test line control valve. A test meter is installed in the test line and another normally closed test line valve installed on the discharge side of the test meter. The pump discharge control valve is closed, both test line control valves are opened, and the pump is started. By controlling the amount of water supply to the pump by gradually opening the control valve on the pump suction, the meter will indicate the flow, and the gauges on the pump will indicate the pressure up through design and overload. The discharge from the metered test line can discharge back into a storage tank, or if no other discharge area is available, it can discharge back into the pump suction.

Some insurance carriers require both the test meter and the discharge header. The discharge header is used to check the calibration on the test meter.

When the test header is used, extreme care must be taken relative to the discharge, which can be the pumps capacity (150%). Hoses can be attached to the 2–1/2″ valves on the test header and run to a safe place for discharge, but the pressures and volumes encountered can cause serious damage if not carefully controlled.

This test header can be used as a hydrant if the location places it in the vicinity of a facility that requires hoses. It should not take the place of regular yard hydrants because of the time required to open valves, but it can play an important role if needed (as a supply for fire department hoses).

There are two types of relief valves: spring operated and pilot operated diaphragm type. The pilot, or pressure, relief valve can maintain a constant system pressure within very close limits as the system demands change. The spring-operated relief valve is set at a predetermined pressure setting, and when this pressure is exceeded, the

Table 5.1 NFPA 20 Fire Pump Data Table

Pump Rating (gpm)	Minimum Pipe Sizes (Nominal) (in.)						
	Suction[a,b,c]	Discharge[a]	Relief Valve	Relief Valve Discharge	Meter Device	Number and -Size of Hose Valves	Hose Header Supply
25	1	1	¾	1	1¼	1 — 1½	1
50	1½	1¼	1¼	1½	2	1 — 1½	1½
100	2	2	1½	2	2½	1 — 2½	2½
150	2½	2½	2	2½	3	1 — 2½	2½
200	3	3	2	2½	3	1 — 2½	2½
250	3½	3	2	2½	3½	1 — 2½	3
300	4	4	2½	3½	3½	1 — 2½	3
400	4	4	3	5	4	2 — 2½	4
450	5	5	3	5	4	2 — 2½	4
500	5	5	3	5	5	2 — 2»	4
750	6	6	4	6	5	3 — 2½	6
1000	8	6	4	8	6	4 — 2½	6
1250	8	8	6	8	6	6 — 2½	8
1500	8	8	6	8	8	6 — 2½	8
2000	10	10	6	10	a	6 — 2½	8
2500	10	10	6	10	8	8 — 2½	10
3000	12	12	8	12	8	12 — 2½	10
3500	12	12	8	12	10	12 — 2½	12
4000	14	12	8	14	10	16 — 2½	12
4500	16	11	8	14	10	16 — 2½	12
5000	16	14	8	14	10	20 — 2½	12

Reprinted with permission from NFPA 20-2016, Standard for the Installation of Stationary Pumps for Fire Protection, Copyright © 2015, National Fire Protection Association, Quincy, MA. This reprinted material is not the complete and official position of the NFPA on the referenced subject, which is represented only by the standard in its entirety which may be obtained through the NFPA website at www.nfpa.org.

valve will open and relieve the pressure on the system. There should be minimal back pressure in the relief valve discharge line so as not to affect the pressure relief with the discharge.

The pipe size in Table 5.1 for relief valve discharge lines must be increased one pipe size if there is more than one elbow in the line, if the line is quite long, or if it rises above the elevation of the relief valve in order to minimize the back pressure. In addition, there cannot be any shut-off valves in the relief valve discharge line. An unofficial rule-of-thumb is not to exceed 15 psi of back pressure in the discharge line of the relief valve. If this level is exceeded, pipe size should be increased. Pipe size increase may be necessary for the pump test line. If the pump test line to a test header is over 15 in length, including the

equivalent length of straight pipe for fittings, the next larger pipe size from that given in Table 5.1 must be used.

If the test line for a test meter arrangement exceeds 100′, including equivalent length of straight pipe added for fittings, the next larger size pipe shall be used, and the meter shall be sized for the increased pipe size. The meter readout device shall be sized according to the rated capacity of the pump. When using a test header, a ball check valve should be installed on the test header side of the control valve before the pipe passes through the wall to the test header. This ball check will automatically drain off condensation and residual water in the line and prevent freezing of the piping through the wall and in the test header.

Valve Supervision

The suction valve must be maintained in its open position. The open position can be assured by installing a device that will activate an alarm when the valve closes. One method of supervision is to install a device that closes or opens an electric contact with the valve wheel. When the wheel turns, the device transmits an audible trouble signal to a constantly attended location. If this electric supervision system is not practical or economical, the suction control valve can be locked open.

Automatic Air Release

An automatic air release valve must be installed on the casing of a split case pump to automatically release air from the pump casing when the pump starts. The automatic air release is required on fire pumps that start automatically. The automatic air release must be a UL listed float-operated type device.

Diesel Fuel System

The fuel supply for a diesel drive fire pump requires at least one gallon per horsepower of the engine, a figure based on one pint per horsepower per hour for 8 h. The fuel supply volume should be verified by the engine manufacturer before specifying the fuel tank. To this volume must be added 5% for expansion and 5% for the fuel that will remain in the sump of the fuel tank. The fuel line to the engine is located so that the fuel in the sump cannot be used by the engine. Note: Where there are multiple diesel drive fire pumps, each engine should have its own dedicated fuel supply tank and fuel supply piping, and the fire pump diesel engine fuel supply should only serve the fire pump diesel engine. The diesel fuel tank should be located in the pump house or pump room, particularly in areas where freezing temperatures are possible. The fuel tank requires a device to determine the fuel

level in the tank. A sight glass type of device is not acceptable.

No shut-off valve can be located in the fuel return line to the fuel supply tank, and the fuel supply line to the engine must be adequately protected against physical damage. The size of the fuel piping and the piping arrangement of the supply and return must comply with the engine manufacturers specifications and requirements.

Pump Rotation

Rotation of the fire pump is important because it determines the location of the suction and discharge, and the pump rotation that is specified or stipulated on the purchase order must comply with the location of the suction and discharge indicated on the pump design. The best way to determine the direction of the pump rotation for a horizontal split case centrifugal fire pump is as follows: Imagine that you are sitting on the (driver) electric motor or diesel engine. If the suction is on your right hand, the pump rotation is clockwise. If the pump suction is on your left hand, the pump rotation is counterclockwise. It is best to design diesel drive fire pumps with a clockwise rotation – the suction on the right. Diesel engines are available with a counterclockwise rotation on special order, but there is usually a delay plus additional cost involved. Ordering the pump with the proper rotation is critical to avoid the delay and expense of returning it.

Engine Exhaust

Each fire pump diesel engine requires a dedicated and independent exhaust system. The exhaust pipe should discharge outside the pump room or pump house, so that the exhaust discharge will not affect personnel or adjacent structures. It is important to check with the engine manufacturer regarding the type of muffler and the size of the exhaust pipe. This is especially true if there is an extensive length of exhaust pipe, to avoid exceeding the back pressure requirements of the specific engine.

Ventilation

The diesel engine manufacturer should be consulted for assurance that adequate air will be supplied to the engine. This critical item is sometimes neglected completely or grossly underestimated in the design of the pump room or pump house. The efficient operation of the diesel engine or engines depends on a supply of adequate makeup air for engine operation.

Diesel Engine Cooling

Two types of diesel engines are available that carry the UL listing and/or the FM approval for use with fire pumps. (Both the fire pump and diesel engine must be UL listed and/or FM approved). These two types are closed circuit, liquid cooled and closed system type.

The first type uses a closed circuit, liquid cooled arrangement with a heat exchanger. This liquid cooling water is taken from the discharge of the fire pump so that any time the pump is operating, water is being supplied to the engine cooling system. With this type of cooling system, the water must be discharged to waste. This means a floor drain is required in the pump room or pump house with adequate capacity to carry off the discharged water without flooding the area. The diesel engine manufacturer can furnish details on the engine cooling system, including the amount of water that will be discharging when the engine is running.

The second type of diesel engine listed for driving a fire pump is a closed system type with a radiator in lieu of a shell and tube heat exchanger. With this type of engine, a floor drain is not required, but adequate ventilation air to cool the radiator must be provided.

Reliable Power Supply to Electric Drive Fire Pumps

Reliability of the electric power supply to the electric motor driving a fire pump is extremely important and requires careful engineering evalu-ation. The NFPA Standard for fire pumps and the FM Data Sheets that describe the requirements for fire pumps both provide detailed requirements for power supplied to electric drive fire pumps. The rule-of-thumb for power supply to electric drive fire pumps is two independent dedicated supplies. When considering a fire pump installation, especially one where the fire pumps are large capacity requiring large horsepower motors, furnishing independent electric power supplies that will meet the requirements of the insurance carrier may not be cost effective. In this case it is best to evaluate the use of diesel in lieu of electric drive fire pumps. Diesel drive election is also worth considering if the fire pumps are larger than the available electric power supply.

Where two fire pumps are required by the insurance carrier-one primary fire pump and one reserve pump – it is sometimes prudent to use an electric drive fire pump for the primary pump and a diesel drive fire pump of equal size for the reserve pump. The use of a primary and a reserve pump is usually a requirement of an insurance carrier when the fire pump is the only source of volume and/or pressure to a suppression system protecting a high value property. The pumps are equally sized for volume and pressure. Should the primary pump fail to start or be out of service for repair, the reserve pump will serve as the primary pump.

Diesel Engine Exhaust

To protect personnel, the engine exhaust pipe should be encased with a high temperature insulation at least 6′ above the floor, and the exhaust pipe passing through a non-combustible roof with a combustible cover should have at least an 8″ clearance between the exhaust pipe and the roof sleeve. If the exhaust pipe exceeds 15′ in length, it should be increased at least one pipe size over that recommended by the engine manufacturer.

Diesel Engine Cooling System

Any discharge from water-cooled diesel engines cooling system should be piped to an open drain located where the discharge is visible. The relief valve waste cone is an acceptable open receptacle for the discharge if the relief valve discharge line discharges to the atmosphere. The diesel engine controller will transmit a signal indicating an overheated engine, low oil pressure, over speed, and failure of the engine to start. If overheating or low oil pressure occurs, the engine will continue to run until it fails. Some controllers have a safety feature to shut the engine down with a low oil pressure or high-water temperature if this condition should occur during the weekly test. If a fire condition should occur calling for the pump start, this shutdown safety feature will be canceled, and the engine will start and run to destruction.

Booster Pumps

When a fire pump is taking suction from a city water main where the volume (gpm) is adequate to satisfy the system demand, but the system requires more pressure than the city main can supply, the fire pump is referred to as a booster pump.

Fire Pumps in a Bypass

When the water supply to a suppression system is furnished by a water source that could provide at least minimum volume and pressure, the fire pump will be installed in a bypass. Should the fire pump be out of service, the system would have a water supply to at least control the combustion.

Pump House

The pump house or pump room should have a reliable heating system to maintain the temperature above 40 °F. The temperature of a pump room or pump house for diesel drive fire pumps should not be less than that recommended by the engine manufacturer. If the temperature in the area where diesel drive fire pumps are installed is not maintained at 70 °F, the use of automatic

engine water heaters or oil heaters is required. These heaters maintain the engine at an efficient starting temperature.

Coupling Guards

The shaft between the pump driver and the pump has a coupling, and since this shaft rotates at a high rate of revolutions per minute when the pump is operating, it is necessary to have a coupling guard to ensure the safety of personnel.

Jockey Pump Controller

The pressure maintenance pump, or jockey pump, has its own dedicated controller which automatically starts the pump when the system pressure drops below a predetermined point, and automatically shuts the pump down when the required system pressure has been restored.

The jockey pump probably derives its name from the fact that it rides on the system. The jockey pump takes suction on the water supply side of the fire pump suction control valve so that it has a water supply even if the suction control valve is closed. The suction side of the jockey pump has an indicating type control valve, and the discharge line from the jockey pump has a check valve and an indicating type control valve. These control valves are intended to facilitate repair of the jockey pump. The discharge of the jockey pump is connected into the system piping on the system side of the fire pump discharge check valve and control valve. Jockey pumps should have very small discharge volume, especially on tight systems (such as sprinkler systems) because there should not be any leakage. If the jockey pump is discharging into large multiple systems supplied by underground piping, the discharge should be sized at least to compensate for possible leakage.

The water pressure in this line reflects the system pressure drop and subsequently trips the pressure switch which, in turn, starts the pump motor or pump engine.

Controller

The controller for diesel or electric drive fire pumps is completely assembled, equipped, wired, and tested at the factory before it arrives on site. For this reason, it is imperative that no alterations or additions be made to the controller after it leaves the factory, as this will void the UL listing and/or FM approval.

Each fire pump shall have its individual control panel, and the control panel should be installed within sight of the pump it controls. The starter switch for an electric motor drive pump or diesel drive engine pump is contained within the control panel. Most fire pumps start automatically, and are stopped manually, with the starter switch actuated by a drop in water pressure. When a suppression system is activated, sprinkler heads fuse, a signal from a detection system trips a valve, and there is a drop in the system pressure as water begins to flow. A small, minimum size water line is connected to the system piping on the discharge side of the check valve in the pump discharge line and piped to the pressure switch in the controller.

Pressure Considerations

The pressure rating of the jockey pump should be 10–15 pounds above the starting pressure of the fire pumps. The fire pumps should never be used as pressure maintenance pumps, and the jockey pump should not have a discharge volume that will satisfy any of the suppression system demands. Fire pumps can also be started on an electric signal from a suppression system, with the system water pressure drop starting as a backup. Automatic starting of a fire pump by activation of a water flow device in lieu of water pressure can be used in situations where the water pressure of the supply fluctuates so much that a predetermined cut-in pressure starting cannot be maintained. Another instance where the pumps may be started by a signal from a detection system or the activation of a water flow device is when the suppression system is protecting a hazardous area and design pressure on the system is needed quickly. The protection of an aircraft from a fuel spill fire is an example.

A few seconds delay can mean the destruction of an aircraft involved in a jet fuel spill fire. The fire pumps, especially the diesel engines that can take a few seconds to come up to speed, should be started to bring demand pressure on the system the second a foam deluge or foam monitor nozzle system commences discharge. When a sprinkler system protects a hazardous area, a fire may almost immediately open numerous heads, and the need for immediate water pressure is imperative. When it is important to start the fire pumps before waiting for the system pressure drop, it is necessary to have the pressure drop starting as a backup.

Controller Cabinet

The controller cabinet, as shown in Fig. 5.3 contains the motor or engine starter, the circuit breaker and the disconnect switch, the manual start device, and the manual stop device. It also provides the relays for alarm and control devices, plus annunciation of the condition of the pump and driver. The controller should be mounted on a dedicated concrete base not less than 12" above the floor to prevent any water from the pumps causing damage to the controller. In designing the location of the controller in the pump room or pump house, the dimensions for the controller being used can be obtained from the manufacturer. If the cabinet is serviced from the rear of the cabinet, allow at least 3-1/2' from the back of the cabinet to the wall, and to permit inspection and servicing, allow not less than 2' on the sides of the cabinet.

Controller Alarm and Signal Devices

Electric drive pump controllers have a power available pilot light on the panel. If the pump room or pump house is not constantly attended, remote audible and visible signals shall be

Fig. 5.3 Fire Pump
Controller and Power
Transfer Switch
(Robert Till)

transmitted to a constantly attended location. These signals are:

- Controller has started the motor and the motor is in a running condition.
- Loss of power to the motor starter.
- Phase reversal on the line side of the motor starter.

Diesel engine drive fire pump controllers have a visible indication that the controller is in an automatic condition. The diesel engine controller should also have the following visible and audible alarms. If the pump room or pump house is not constantly attended, these signals should be transmitted to a constantly attended location:

- Low oil pressure
- High engine jacket coolant temperature
- Failure of engine to start automatically
- Engine shutdown from overspeed
- Battery failure
- Battery charger failure

Some controllers are required to have a relay that will start the pump motor or pump engine when there is a power failure to the controller. With electric motor drive pumps, if the power supply is adequate, the motor should start across-the-line to full power on signal from the starter. If the power supply is not capable of accepting the surge of power, the motor can use a reduced voltage starting sequence where the motor comes up to full speed slowly or in stages. When using multiple electric drive pumps, the motors should be started in sequence, with a several second interval between the starting of successive motors to prevent a breakdown of the power source because of the heavy load.

Weekly Diesel Drive Pump Test

The controller for a diesel drive fire pump must be equipped to provide an automatic start of the engine once a week, and a device to keep the engine running for not less than 30 min. The controller can be equipped with an automatic shutdown. This weekly test run can be recorded on a pressure recorder.

Fire Pump Controller Alarm Signal Operations

The following options can be provided as relays in the controller to transmit trouble signals to a constantly attended location. Some of the following may be required by the insurance carrier:

- Low pump room (house) temperature
- Relief valve discharging
- Flow meter left on by passing pump
- Suction supply water level below normal
- Suction supply water level near depletion
- Diesel fuel supply below normal

Diesel Engine Batteries

The diesel pump starts from not less than two batteries that are mounted on the engine base. The batteries are constantly charged by a battery charger that is specifically listed for use with fire pumps.

Vertical Turbine Fire Pumps

When it is not possible to have a water supply to the pump under a positive pressure, the use of a vertical turbine fire pump should be considered. A vertical shaft pump is a centrifugal pump with the impellers in a column. The primary consideration with vertical fire pumps is that the source of water is adequate and dependable. This source can be from reservoirs, lakes, or rivers supplying wet pits. Vertical turbine fire pumps can be driven by an electric motor mounted on the top of the pump shaft, or by a diesel engine operating the pump shaft through a right-angle gear. The design of the supply to a vertical turbine fire pump is rather sophisticated and should

be calculated and designed by experienced fire protection engineers. The support of the pump shaft and electric motor or right-angle gear drive is a consideration of a professional structural engineer. The pump house design must incorporate a closable opening in the roof to allow the installation and removal of the pump shaft. The electric motor for a vertical pump must be equipped with non-reverse ratchets, and the right-angle gear used when the pump is driven by a diesel engine must be equipped with a non-reverse ratchet (Fig. 5.4).

Flushing and Testing

Before performing the fire pump acceptance tests, the piping must be hydrostatically tested at 200 psi for 2 h with no leaks. If the system pressure is to be greater that 150 psi, it will be necessary to determine the maximum system pressure and establish the hydrostatic test pressure at 50 psi over the maximum system pressure. The fire pump suction piping must be flushed at a predetermined flow rate. Once the system demand has been determined by hydraulic calculations, the flushing flow rate shall not be less than the calculated system flow rate. It is possible that the calculated system flow rate will not be sufficient to meet the flushing flow rate requirements of the NFPA, which are based on the pipe size of the suction piping.

If the pump has a 10″ suction pipe and the calculated system demand is 1800 gpm, a check of the NFPA Standard for fire pumps, Standard 20, indicates that the flushing flow rate for a pump with a 10″ suction is 2440 gpm. NFPA 20 will indicate the greater flow rate system demand gpm or NFPA flow rate requirements. When the flushing and testing has been completed, the contractor will provide a certificate verifying the successful completion of the flushing and testing.

Fire Pump Field Acceptance Test

When the tire pump installation is subjected to the field acceptance tests, the following representatives must be present:

Fig. 5.4 Vertical
Turbine Pump.
(Courtesy of AC Fire
Pump)

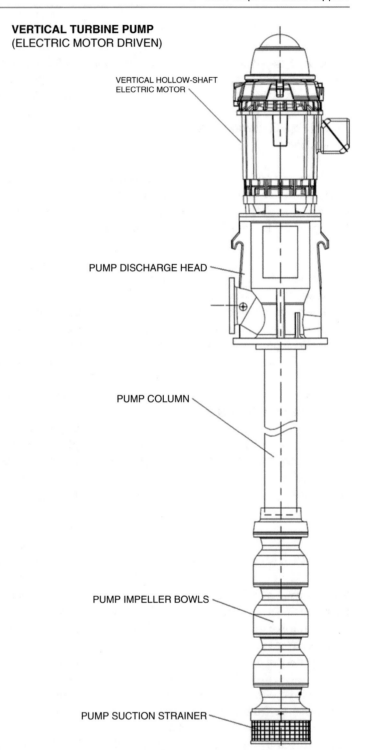

VERTICAL TURBINE PUMP
(ELECTRIC MOTOR DRIVEN)

VERTICAL HOLLOW-SHAFT
ELECTRIC MOTOR

PUMP DISCHARGE HEAD

PUMP COLUMN

PUMP IMPELLER BOWLS

PUMP SUCTION STRAINER

- Pump manufacturer
- Engine manufacturer (if diesel drive)
- Controller manufacturer
- Transfer switch manufacturer (if supplied)

These representatives include the cost of
attendance in the quotation or at least have some
agreed financial arrangement with the contractor.
The owner, insurance carrier, engineer, and all

authorities having jurisdiction (or their appointed representatives) should be present during the tests. Parties should be notified of the place, date, and time of the tests not less than 5 days before the test date.

There is always a concern that because the pump water supply is not adequate to allow testing the fire pump at overload, or 150% of its design, the test will not be successful. If this situation exists, the pump shall only be required to discharge the maximum that the supply is capable of furnishing to complete this phase of the test successfully. The installing contractor will furnish all labor, material, and devices required for the tests. It is recommended that the contractor run the pump through the test procedure a few days before the official field acceptance test date, to uncover any deficiencies and make the necessary corrections or repairs. The fire pump will be required to discharge at the minimum, the rated, and the peak pump loads, and the results will be compared with the manufacturers certified pump test curves. The controller manufacturer has a recommended test procedure, but it is important that this procedure include a minimum of 10 automatic and 10 manual pump starts, and that the pump driver operate at full speed for not less than 5 min during each of these test starts. All alarms, both local and remote, will also be activated and tested.

Following completion of the acceptance tests, the fire pump installation should be returned to the automatic operating condition, including refilling of the diesel fuel tanks. The NFPA Standard 20 and the FM Data Sheets contain information regarding field acceptance tests. In addition, the insurance carrier may have specific acceptance test requirements. It is recommended that the insurance carrier input be obtained relative to their acceptance test requirements. The preparation of a test schedule will expedite the test procedure, especially if the acceptance test involves a multiple pump installation.

Miscellaneous

When pump manufacturers submit quotations for a fire pump project, the difference in costs may be a reflection of what is included in the quote. It is important to stipulate in the bid request exactly what equipment should be included in the quotation. The following is a list of items the pump manufacturer can and should furnish:

- Pump
- Driver
- Controller
- Jockey pump and jockey pump controller
- Flow meter easer
- Hose test header
- Relief valve
- Closed waste cone
- Open waste cone
- Circulation relief valve
- Eccentric reducer
- Concentric increaser
- Pressure recorder
- Relief valve tee

Water Supplies

Bulk supplies of water are necessary to put out any fire. When a free-flowing water supply is not available, water must be stored locally in tanks and reservoirs. When an adequate supply of water is available, but the pressure is not adequate to move the water to the fire, a fire pump must be employed.

Tank Selection

Water storage for fire protection is available in ground level tanks, elevated storage tanks, and ground level reservoirs. The ground level steel tanks are referred to as pump suction tanks and are available in a wide range of capacities. Approval of the fire tank system must be obtained from the insurance carrier. Factory Mutual has a list of approved tank designs, capacities, and dimensions, and the data sheets mandate the construction and installation of these tanks. NFPA 22 – Water Tanks for Private Fire Protection-mandates the construction, design, and installation of water storage facilities for fire protection system water supplies.

If the client is insured by Factory Mutual, the system must use FM approved tanks and installed in accordance with FM standards. If the client is insured by an insurance carrier other than FM, the specifications, cost estimates, and purchase orders must stipulate that the tanks conform to the requirements of NFPA relative to construction, design, and installation.

To lend some flexibility to these statements, a manufacturer of Factory Mutual approved water storage tanks has the option to design, fabricate, and erect a tank using dimensions other than those that have been approved by FM, but the tank is subject to review and approval by Factory Mutual Research. FM Research is part of the FM insurance family that establishes the specifications for FM approved devices, materials, systems, and equipment, and lists them in the FM Approval Guide. In the Factory Mutual Approval Guide, the dimensions-both height and diameter-for each tank capacity are presented. Factory Mutual also lists approved elevated gravity tanks in standard sizes.

NFPA 22 also details dimensions and capacities for a variety of tanks, including wooden elevated gravity tanks or wooden towers. Although some yard or site conditions dictate the dimensions of a ground level pump suction tank, or the height, dimensions, and location (possibly on a building in lieu of a tower) of an elevated gravity tank, if at all possible the use of an FM approved standard tank design and size will save the trouble, time, and expense of obtaining approval of a special design. In geographic areas subject to freezing, these tanks require an approved form of freeze prevention.

Concrete Reservoir

If space is available, a concrete ground level reservoir can be provided for fire protection water. These reservoirs come in all sizes and shapes, covered or open. The suction originates from a point in the reservoir that provides a positive suction pressure to horizontal pumps.

The pit where vertical turbine pumps are employed must be designed very carefully to allow adequate room around the pump shaft to prevent the water flow from forming a vortex, which could affect the efficiency of the pump. Webster defines a vortex as a "whirling mass of water which forms a vacuum at its center." It is easy to visualize the affect on the efficiency of the pump with the pump shaft in the center of a vortex. A mini-vortex is the whirling water rotating around the open throat of a drain. In a wet pit, the pump shaft with the impellers is rotating at tremendous revolutions per minute. The pump manufacturer should provide the correct dimensions of the wet pit for their specific vertical turbine pump.

In pump suction tanks, the suction pipe is installed near the bottom of the tank, with an elbow on the end of the suction pipe in the tank. To avoid the creation of a vortex as the water rushes into the suction pipe elbow, a vortex plate is fastened to the elbow.

The vortex plate fastened to the suction elbow is a 4′ by 4′ square plate, located a distance one half the diameter of the suction pipe above the tank floor, but not less than 6″ above the tank floor. Water entering the suction elbow must pass around the square sides of the vortex plate, which prevents the swirling or whirling action of the water entering the elbow.

A single bare end suction pipe would create the same action as the water entering the open drain, accented because this water is being sucked into the suction pipe in a great capacity and with considerable velocity, and a vortex would surely form. Where yard site space was limited, concrete reservoirs have been constructed under the floor of the pump room or beneath the basement floor of a building, using vertical turbine pumps to supply water to the suppression systems. It was previously noted that two above ground steel pump suction tanks were required to be installed for reliability of the water supply. With concrete reservoirs, it is only necessary to split the reservoir into two sections for the same results. If one side of the reservoir should be drained for repair or cleaning, the other side of the reservoir would still have an adequate water supply to meet the system demand. The piping supplying the pumps from this divided reservoir must take suction individually from both sides of the reservoir with control valves in each suction pipe, and the two pipes connected to a common pump suction header. This design allows the pumps to take suction from either side of the reservoir, or both sides simultaneously.

Embankment Fabric Reservoir

A reservoir that deserves serious consideration as a fire protection water supply is an embankment-supported, rubber-coated, fabric tank. The tank is actually a reservoir liner with a cover. When the liner is filled with water, its top floats upward and remains floating when the tank is full of water. The earth banks are built on all four sides of a square, and the tank liner is laid, unrolled, in the square within the embankments. Once the liner is in place within the open enclosure, it is anchored to the embankments and the piping connections are completed. The tank is constructed of a high-strength nylon fabric that is coated with synthetic rubber that not only seals the liner watertight, but also protects it from deterioration from weather, abrasion, temperature changes, chemical fumes, and ozone. The synthetic rubber coating keeps the liner flexible and shock-resistant, and therefore the tank is able to resist severe damage in earthquake areas. The inlet-outlet drain fitting is installed in a concrete pad with a vortex plate bolted to the inlet-outlet drain fitting. The exposed surface of the filled tank is painted as recommended by the manufacturer and is applied according to the recommendations. In areas subject to freezing, the tank water should be heated by circulating the water through a heat exchanger.

Tank Heating

Embankment reservoirs, like all the tanks and reservoirs described, must be constructed, installed, and heated (in areas subject to freezing), in accordance with the requirements of NFPA 22 and the Factory Mutual Data Sheets. The map in these publications indicates (by means of temperature lines) the lowest 1 day mean temperature for all areas of the country. By locating the tank site on this map, the lowest temperature that the tank will experience establishes the amount of heat that will be required. This determination of the lowest 1 day temperature the tank will experience, and the capacity of the tank, are both factors that are used in calculating the amount of heat necessary to maintain the water temperature in the tank at or above 42 °F. A low water temperature alarm should be provided to alert personnel if the heating system should fail to maintain the minimum temperature of the water. The heating arrangement should automatically control the water temperature in the tank.

Tank Water Level

Tanks and reservoirs should be kept full automatically by use of an altitude valve arrangement or some other approved configuration. All tanks should have some device to determine their water level, and a low water level alarm device should the water level fall below normal capacity.

Summary

Tank and reservoir designs should be engineered by experienced professionals and involve all concerned disciplines-civil, structural, mechanical, and electrical. The location of the tank requires the expertise of a civil engineer, the design of foundations and structure of a concrete reservoir mandates nothing less than the engineering expertise of a structural engineer. The piping and heating equipment, design and engineering, should only be trusted to a mechanical engineer, and all alarms and control devices and equipment can only be engineered and designed by an electrical engineer. The fire protection engineer can determine the required capacity of the tank, and select the appropriate, cost effective, water storage unit. He or she can design and engineer, and specify the required fire protection features, all based on experience and expertise and to meet the requirements of NFPA and/or FM and the insurance interest having jurisdiction.

Underground Fire Mains

6

Underground Fire Mains

Selection of underground pipe for fire protection depends on several considerations: an evaluation of the soil conditions, the maximum working pressure, and the external loads the pipe may experience by virtue of vehicular and other traffic. Fire protection water mains must be capable of withstanding the unusual, and sometimes severe, conditions associated with fire protection systems:

- fire pump pressure surges
- hose stream operation
- deluge system operation
- the sudden closing or opening of valves

Cast iron, ductile iron, steel, and asbestos cement underground piping do not carry either UL listing or FM approval. This pipe, when used for underground fire mains, is acceptable to Factory Mutual and to the insurance carriers who normally rely on UL listed equipment, as long it is manufactured according to the standards of the American Water Works Association (AWWA) and the specifications of other nationally recognized engineering organizations.

This chapter explains underground pipe selection, hydrant selection, flushing, and testing for the various types of piping listed above.

Underground Pipe Selection

Suppression systems that use water as their extinguishing agent are usually designed for a maximum working pressure of 175 psi. If the system pressure is expected to exceed 175 psi, extra heavy fittings (250 lb, 300 lb), valves, and devices must be used. In addition, the sprinkler heads must be UL listed, another factor that adds to the system installation cost. Systems that will not exceed this pressure limit can use standard weight fittings, valves, and sprinklers. System pressure levels also have a design impact on both the underground fire mains and the above ground system. For example, pump installation requires considerable experience to balance pump shut off pressure of 120% of the rated head for vertical turbine pump suction and system demand, while maintaining the system working pressure below 175 psi. Shut off pressure can be particularly difficult while attempting to size a fire pump to deliver the system demand without exceeding the 175 psi level.

Cast Iron Pipe

For standard systems (working pressure below 175 psi) underground cast iron and ductile iron pipe rated as Class 150 is most commonly specified.

© Springer International Publishing AG, part of Springer Nature 2019
R. C. Till, J. W. Coon, *Fire Protection*, https://doi.org/10.1007/978-3-319-90844-1_6

The Class 150 rating represents the main working pressure. This distinction can be confusing, as compared with the 175 number. The 150 psi classification for cast iron and ductile iron pipe (and most flanged valves) is the maximum working pressure with high temperature fluids. At the normal operating temperatures of suppression systems, namely transporting water, the 150 psi working pressure is adequate to handle pressures in excess of the 175 psi. These ratings are normally cast into the valve bodies: For example, 125 WSP (working steam pressure), and 200 WOG (water, oil, gas). For 175 psi suppression systems and the laying conditions previously described (namely 5′ bury and a tamped backfill) the pipe should be listed as Class 22, 150 pound working pressure. If system pressure will exceed 175 psi, cast iron underground pipe Class 200, 250 and up should be considered based on an evaluation of all of the conditions involved.

Ductile Iron Pipe

This type of piping is, as its name implies, much more flexible than cast iron. Ductile iron will bend considerably before it will ultimately fail and exhibits a tremendous impact strength. One manufacturer of ductile pipe demonstrated its strength by dropping a length of pipe from a great height onto a rail without having the pipe length break. In another demonstration, a 4500 pound weight was dropped on a length of ductile iron pipe, and the pipe was flattened but did not break. The strength of ductile iron pipe makes it desirable for use under heavy vehicular traffic conditions and in unstable soil where the pipe could be subjected to the stress of traffic loads. The tremendous pressure impact of a water hammer has no effect on ductile iron pipe, where cast iron might crack or rupture.

Specifics for Cast Iron and Ductile Pipe

Cast iron and ductile iron pipe for fire protection should be lined pipe (cement/enamel/polylined), to prevent the buildup of tuberculation within the pipe interior. This type of growth is shown in Fig. 6.1. Tuberculation can be a precursor or occur with MIC (Microbiologically Influenced Corrosion).

Cast iron and ductile iron pipe joints can be bell and spigot, mechanical, and push-on type. The spigot, or plain end of the pipe, is inserted into the bell end of the pipe; packing, usually jute, is fitted into the bell around the spigot end; and the joint is sealed with molten lead. Although bell and spigot use to be the most common type of cast iron pipe joint, they require considerable labor to assemble. As a result, bell and spigot joints are seldom used in fire protection mains in todays labor conscious market. Mechanical joint pipe push-on joints are basically a modified bell and spigot arrangement where the plain end of the pipe is inserted-pushed-into the bell end over a rubber or neoprene 0 – ring type gasket, compressing it, thereby sealing the joint.

Friction Loss: The "C" Factor

The coefficient, or constant, for friction loss in underground piping is represented by the letter C, followed by an indicating number. The larger the number following the letter C, the less friction loss in the pipe. For example:

- Cast iron unlined-new C = 120
- Cast iron lined-new C = 140
- Cast iron in use (public or industrial main) for moderately corrosive water:

Fig. 6.1 Tuburculated pipe. (Courtesy of Fred Hart)

- 10 years old C = 90
- 20 years old C = 65
- 30 years old C = 55

These C factors can make a major difference in pressure loss when hydraulically calculating the underground piping supply to Are protection systems. For example:

- 1000 gpm, 6″ cast iron pipe, C = 100, friction loss per 1000′ of pipe is 52 psi.
- 1000 gpm, 6″ cast iron pipe, C = 140, friction loss per 1000′ of pipe is 27.92 psi.

Other Types of Underground Piping

There are some instances in which soil and/or water conditions can cause considerable corrosion with cast iron or ductile iron pipe. The water in the soil may be extremely corrosive to these pipe materials, because of the acidity and alkalinity of the area. Through a complicated electrochemical reaction, external corrosion of cast iron and steel pipe can be relatively rapid. Stray currents can follow the buried piping to a point where the soil has less electrical resistance than the pipe, and the current leaves the pipe. This creates ionization that is similar to soil corrosion. When a soil survey reveals these electrical currents from an external source, there are two methods of preventing this current from causing severe erosion of the cast iron pipe-namely coating and wrapping the pipe or using cathodic protection. Currents can be bled-off by providing low resistance metallic grounding connections. Basically, cathodic protection applies a direct electric current to the pipe line to nullify the stray current.

Polyvinyl Chloride Plastic Pipe (PVC)

A reliable substitute for cast or ductile iron pipe, polyvinyl chloride (PVC) piping is highly resistant to corrosion, is available in Class 150, and is lighter in weight. In fact, its relatively light weight provides a considerable labor savings cost impact due to ease of handling and installation. This comparison between the additional material cost of PVC

over the material cost of handling and installation and the decreased cost of installation for PVC over cast iron, must be carefully evaluated if cost is the primary reason for using PVC pipe. PVC can be field cut with a power saw and uses a modified bell and spigot type joint with a ring placed in a groove in the bell end that provides a tight seal when the spigot end is installed in the bell end.

Flanges can be attached to the PVC to accommodate attachment to cast iron valves or fittings. Being nonmetallic pipe, PVC is not subject to external or internal corrosion, nor will it experience the effect of electrical currents that cause external corrosion. The C factor for PVC is 150, compared with new lined cast iron and ductile iron pipe with a C factor of 140. PVC underground pipe may be required due to soil conditions that would corrode other piping materials, or to take advantage of the improved C factor of 150. If there is a considerable amount of underground fire main involved, the use of PVC could mean the difference in hydraulically calculating the demand between a large cast iron or ductile iron pipe size, and a smaller PVC fire main. (See the chapter Hydraulic Calculations of Sprinkler Systems for an explanation of hydraulic calculations.) It should be noted that PVC are not as strong as metals in terms of burst pressure and collapse pressure as pipe diameters increase.

Fiberglass-Reinforced Plastic Pipe

Another nonmetallic underground pipe which resists corrosion, fiberglass-reinforced pipe, also does not require cathodic protection. Like PVC, this pipe is lighter than cast or ductile iron, and the joints are of a modified bell and spigot type, where the spigot end is coated with an adhesive before being inserted into the bell end.

Asbestos Cement Pipe

This pipe is another material that does not corrode, and usually consists of a sleeve (coupling) with an 0-ring gasket at each end. The ends of the pipe are tapered, and when inserted into the sleeve they compress the gaskets which make a tight joint. The C factor for asbestos cement pipe is 140.

Due to its nature as a carcinogen, asbestos cement pipe is no longer used, and a lot of it needs to be replaced because of its age.

Steel Pipe

Steel pipe can be used for buried fire mains using welded joints with the pipe coated and wrapped, but even with this corrosion precaution, the corrosion factor of buried steel pipe is a major concern. If it ever becomes necessary to bury a dry pipe system, the use of steel pipe with welded fittings and joints is the only means of retaining the compressed air without excessive loss. Fire mains smaller than 6″ may be used, but they must be lined, their use must be proven by hydraulic calculations, and the main must be as large as the system riser. The cost of installing a 6″ fire main is almost equal to installing a 4″ main. The material difference is very minor, and the trench work almost identical, so always consider the advantages of a 6″ supply to that of a 4″ in relation to the slight cost increase. The bury or cover of earth over the top of the pipe, depends on the temperature of the area (which determines the depth of the frost line). Maps are available which indicate the recommended bury depth of piping throughout the country determined by the maximum depth of frost penetration. In locales where frost is not a consideration, underground piping should not have less than 2-1/2′ of bury to protect the piping from mechanical damage. Where freezing is a factor, depth of static or stationary fire protection piping should be 6″ deeper than municipal water mains which have a regular flow of water.

Underground Pipe Bury

The term bury is used to establish the cover on the pipe. When determining the required depth of the pipe trench, be certain to provide the required bury while compensating for site grading. When underground piping passes under a building foundation, the bury should be the same as that required for outside bury – a sleeve through the foundation wall is the most common solution. There are times when it is absolutely necessary to run the fire main under a footing that is at greater depth than the pipe line. When this occurs, the pipe must not be raised to pass above the footing, otherwise it will not have the required bury. Extreme caution must be taken in fill compaction in the area to prevent future settling which can shear through the pipe line. This condition mandates other innovative solutions, which may prove to be more expensive; for example, trenching at a greater depth for the entire run of piping or placing 45° ells to drop the line under the footing. However, violating the required bury leaves open the potential of the water laying static in the pipe freezing, and subsequently cracking the pipe, which is much more expensive than the innovative installation solution. Even if the freezing is not severe enough to crack the pipe, any ice formation will obstruct the water flow that the suppression system needs in a fire.

Rodding and Thrust Blocks

Tremendous thrusts are created in the buried piping wherever the pipe changes direction, due to pressure surges in fire protection underground piping when flow is started or stopped. Most cast iron and ductile iron pipe, bell and spigot, or mechanical joints, will use rods that run from fitting to fitting, or fitting to joint to hold the joints, fittings, and valves in opposition to these thrusts and relieve the pressure that could cause a joint to separate.

When soil conditions are not stable enough to hold a fitting in place when a water surge occurs, it is necessary to install thrust blocks. Under any number of conditions – excessive pressure in the main, unstable soil-thrust blocks may be required, even if the piping is rodded. This of course requires a thorough engineering evaluation.

Thrust blocks are masses of poured concrete, poured against the fitting where the pipe changes direction, or against a valve that could be subject to surge pressure. The consistency of the concrete, the size of the thrust block, and the technique for pouring the thrust block must be in accordance with the authority having jurisdiction. It is also this authority, especially if they are an insurance carrier, who may dictate the use of thrust blocks, rodding, and retainer glands (devices used on mechanical joint pipe to prevent separation of joints). Whether or not the stability of the soil is known, it is always advisable to meet the insurance requirements.

Bell and spigot pipe can be rodded by installing a pipe clamp and rodding through the pipe clamps. But in the majority of cases there is no question that nonmetallic pipe like PVC must have thrust blocks. There are very few instances where the use of thrust blocks will not be required when using piping other than mechanical joint cast iron or ductile iron piping.

Other Considerations

Underground piping must never be installed below the floor of a building. This is against NFPA Fire Codes and good engineering practice. Installing piping below the floor means that a leak or break would not be detectable, and could undermine the floor, causing collapse. Even if the leak or break were detected through a meter, finding the location of the leak and repair would be a costly operation. When there is an underground piping system that supplies several hydrants and/or several suppression systems, or even several buildings in an industrial or office complex, the fire main should be sectionalized. This means installing post indicator valves (PIVs) at various locations so that, should a repair of the fire main be required, the entire fire main supply would not have to be shut down-the section between two sectional PIV s could be isolated by closing just these two valves. There is a rule of thumb that sectional valves should not control more than five systems, with hydrants being considered as a system. This rule exists because the placement of sectional valves must be reviewed and approved by the insurance interest having jurisdiction.

Hydrants

The installation of fire hydrants supplied by a private fire main involves three types of hydrants. One is a wet barrel hydrant used in areas that are not subject to freezing because the water fills the hydrant exposed above ground.

The second type of hydrant is a dry barrel hydrant where the water is kept at its base by a valve that is opened when the hydrant nut is operated.

The third type is the wall hydrant. It is installed on the wall of the building, is similar in appearance to the fire department siamese connection, and water is controlled by a non-rising stem valve located inside the building with the operating nut on the exterior wall. In most cases, the hydrants that are installed on private property supplied by private fire mains consist of 2-1/2″ hose outlets. Most public hydrants also have a larger, usually 4″, pumper connection, but insurance carriers may not permit the pumper connection on private hydrants. They are usually concerned that too much water is being taken from this outlet, thereby decreasing the supply to the suppression systems. To cite an actual project experience:

> The hydrants for installation on a private property fire main were incorrectly ordered with the 4″ pumper connection. The insurance carrier required the owner to weld the 4 outlets shut so there was no chance they would be used.
>
> Private hydrants should be located approximately 300–500′ apart, using the 300′ spacing in areas of greatest hazard, and 500′ between hydrants in other areas, but no hydrant hose length should exceed 500′. Placement of hydrants becomes an experienced engineering judgment, and here again, their location is subject to the authority having jurisdiction.

Flushing and Testing

Above-ground suppression systems must be hydraulically tested, but underground fire mains must be flushed and tested. The piping must be flushed at the water demand rate of the system, and this demand flow must be at least of a capacity to produce a velocity in the flushing water of 10′

per second. To produce this velocity, the following gpm must be discharged through the appropriate pipe size:

4″ pipe 390 gpm
6″ pipe 880 gpm
8″ pipe 1560 gpm
10″ pipe 2440 gpm
12″ pipe 3520 gpm

After the piping is installed, and before conducting the hydrostatic test, the installing contractor should not backfill and cover the piping joints or valve connections. However, it is mandatory to backfill the trench between the joints to prevent any pipeline movement during the pressure test.

All flushing and tests must be witnessed by the authorities having jurisdiction, the owner, or appointed representatives, and at the conclusion of the flushing and tests, the installing contractor must furnish a signed Contractors Material and Test Certificate, countersigned by the witnesses. If the installing contractor is not performing the tests, the above ground system contractor must receive a copy of the flushing and test certificate. The above ground contractor is assured that the main has been flushed and tested, and any problems with obstructions in the above ground system are his responsibility.

To complete the testing of the underground piping system, each hydrant supplied by the fire main must be opened and closed while the system is under pressure to uncover any hydrant leaks. In addition, all control valves on the underground fire main must be closed and opened while the system is under operating pressure to ensure that the valves are opening and closing properly.

Equipment and Devices

7

Equipment and Devices

The proper location of valves and other critical devices is important to ensure that the system may be shut off promptly after sprinklers have extinguished the fire, or to conserve water and pressure if pipes are broken. Equipment such as strainers are necessary to prevent clogging of the nozzles. These devices, as well as backflow preventers, tap connections, and hangers, are integral parts of any fire prevention system. The purpose, use, and installation of these important elements are discussed in this chapter.

Control Valves

Every suppression system must have a control valve, with one requirement-the valve must visibly indicate whether they are open or closed.

The Outside Screw and Yoke

The OS&Y valve is one of the most common and easily recognizable fire protection system control valves (Fig. 7.1). This valve has a threaded stem that projects several inches above the operating wheel when in the open condition. When closed, the valves threaded stem retreats into its yoke, and the top of the stem becomes flush with the wheel; thus, its name, OS&Y is derived from its operation. The outside screw rises above the yoke when it is open, and the yoke is part of the body of the valve where the stem lodges when the valve is closed.

The Post Indicator Valve (PIV)

This is one of the most common fire protection valves, controlling a system from outside the building. The post indicator portion of the valve is a post that protrudes approximately 3′ from the ground with a window-an open slot in which the word open or shut appears (See Fig. 7.2). A cast steel wrench is attached to the top of the post that is designed to house a padlock for security. A wrench fits the nut on the top of the valve stem, and when the stem is turned to dose the valve, the window will read shut.

When the PIV is installed on the underground piping, the valve part is a non-rising stem type valve (N.R.S.). This means that when the valve is opened, the stem does not rise vertically as it does in the OS&Y valve. When the valve is opened or closed, the stem turns within the post indicator portion of the PIV.

When PIV valves control the water supply to a suppression system, they serve two main functions. First, the control valve is closed when making repairs on the system or when remodeling or adding to the system. The second primary function of the control valve is to shut the system down in a fire situation. For example, a fire has gotten out of

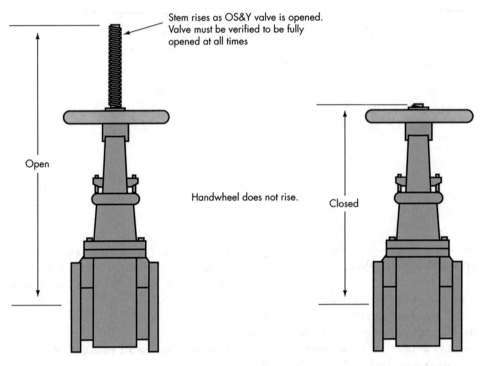

Stem rises as OS&Y valve is opened.
Valve must be verified to be fully
opened at all times

Open

Handwheel does not rise.

Closed

Fig. 7.1 OS & Y Valve. (Courtesy of John Morrison)

Fig. 7.2 Post Indicator
Valve (PIV). Photograph
by Robert Till

control and opened so many sprinkler heads that the water supply pressure at the hydrants has dropped below the fire departments requirements. Another example might be that a roof may collapse due to the effect of heat from a fire on the structural steel; the collapse may break a large sprinkler supply pipe; or an explosion destroys the sprinkler supply piping.

Any of these incidents would create a tremendous loss of fire protection water not directed toward extinguishing the fire. This loss causes decreased pressure and volume at the fire hydrants. That seriously hinders firefighting. If the control valve on the system is an OS&Y valve located inside the building, in all probability it would not be accessible. A similar situation may occur if the fire approaches the system riser and control valve location, because the fire could prevent access to the control valve. This is the primary reason that the authority having jurisdiction over approval of fire protection plans and installation will require a PIV, located outside the building for control of the system water supply.

The ideal location of the PIV is about 40′ from the building, to allow access to the control valve even if a fire breaks out of the exterior wall or an exterior wall collapses. Despite this requirement, there are many situations where it is impossible to locate the PIV 40′ from the building, or where use of a PIV is not possible (when there is no yard around the building, or the entire area is congested with roads).

There is an alternative-the use of a wall type post indicator valve. The wall type indicator post valves post is installed on the exterior of the building and the valve shaft passes through the wall, connected to the non-rising stem valve installed in the system riser.

There are a number of types of wall posts operated with a wheel in lieu of a wrench, and all have the window shut/open indicator. One type of wall post extends about 14″ from the face of the wall, one type has a post extending up to 52″ from the wall; and the third is completely recessed in the wall in a cast iron frame.

Important Considerations When Using the PIV

When ordering the post indicator valve for underground piping, the distance from the ground level to the bottom of the main is important. Naturally it is imperative to state the size of the main to obtain the proper size valve, but the depth of the underground pipe trench, which is the bottom of the pipe, is also critical because the post extension that connects the post to the valve is available in various lengths. This dimension must be fairly accurate, or the post will either protrude too far above ground (unsightly) or too close to the ground (unseen). The rotation of the valve opening is another consideration-shall the valve open by turning right or left, clockwise or counterclockwise? The new valve should open the same way that other valves on the property open, in order to avoid confusion in a fire condition. Should a valve be forcefully turned the wrong way, it could damage or possibly snap the valve stem. Unless otherwise specified, the PIV will turn in a counterclockwise opening direction.

For wall type post indicator valves, it is important to designate the center line of the riser where the non-rising stem valve is located.

Butterfly Valve

The indicating butterfly control valve is a popular substitute for the large OS&Y valve. This is a thin valve which does not have the long yoke and stem of the OS&Y valve. Therefore, it does not take up valuable installation space both in and around the perimeter of the system riser.

For example:

Installed in a system riser as the control valve, the flanged butterfly valve takes out 5″ for a 6″ riser, and 6″ for an 8″ riser. By comparison, a flanged OS&Y valve takes out 10–1/2″ when installed in a 6″ riser, and 11–1/2″ in an 8″ riser. The center line to the outside of the handle of a butterfly valve measures approximately 11–1/2″ for a 6" valve, and 12–1/2″ for an 8″ valve. The 6″ OS&Y valve, center line to the tip of the stem in the open position, requires almost 32″ for a 6″ valve, and almost 41″ for an 8″ valve.

The example compares flanged butterfly vales with flanged OS&Y valves, but when the butterfly valve is designed to be used with grooved couplings or fittings, the thickness is much less than that of the flanged butterfly valve. In order to qualify as an indicating type valve, the butterfly valve has an arrow indicating the open or shut condition. The butterfly valve is available with either a wheel or crank operator and can be provided with supervisory switches to signal a trouble alarm should the valve be closed. The physical dimensions and weight difference also make the butterfly valve much easier to install, cutting down on labor cost.

Pit OS&Y and Post Indicator

If a building or facility has multiple fire suppression systems, and they are all supplied by a common below ground fire main, it is good practice to provide a single valve pit in the main line that feeds the fire suppression systems. This arrangement is usually a requirement of the code authority to prevent the fire department pumper from being connected to the wrong building siamese connection, rather than to the building needing protection. The valve pit has a yard type fire department pumper connection that sits on top of the concrete pit. When the fire department pumps into this siamese, they are pumping into the entire underground piping system and into whichever building system is operating. The concrete valve pit contains:

- an OS&Y valve
- a check valve to prevent the fire department from pumping back through the system water service
- a second OS&Y valve
- the tee for the fire department pumper connection (which also has a check valve in the line to the siamese).

If a post indicator valve is required underground, it is possible to use a non-rising stem valve in lieu of the second OS&Y valve and, with an extension up through the top of the concrete pit, install a post indicator mounted there. Valves are necessary on either side of the check valve so that they could be closed, and the check valve removed for repair or cleaning. With the connection for the fire department pumper on the house side of the control valve, when both valves are closed, and the check valve removed, the fire department may still use the pumper connection to supply water under pressure to the suppression systems.

Check Valves

Underwriters Laboratories and/or Factory Mutual-approved swing check valves provide a minimum of friction loss. The check valve, disc, or clapper swings entirely clear of the flow of water. There is now available a compact wafer check valve. This is smaller than the iron body check valve. For example:

> An iron body check valve measures 17″ face to face of the flanges, and an 8″ iron body check valve measures almost 21″ for the same. By comparison, a 6″ wafer check with grooved ends measures a little over 9″ end to end. The disc on the larger iron body check valves is usually cast iron, and the wafer check valves usually have spring loaded bronze discs.

Wafer check valves are not only compact but weigh much less than the conventional swing check and, therefore, are labor saving as well as materially cost effective.

Detector Check Valves and Full Flow Fire Meters

Some cities require that the fire water supply to suppression systems be metered. Originally this was to keep citizens from stealing water from the unmetered fire supply. Although extremely difficult without operating all the alarms or energizing the fire pumps, this theft has occurred. For cities that do require metering, the cost of water used to fight a fire is not their concern because as much, if not more, water would be used by the fire department using the unmetered fire hydrants. Nonfire usage should be metered.

It was out of this concern that the detector check valve was developed. This valve is basically a regular iron body swinging disc check valve, with the disc, or clapper, weighted. Tapped into each side of the check valve around the disc is a small pipeline with a water meter. When there are small flows in the fire main, the weighted clapper will not rise off its seat and the water flow, seeking the line of least resistance, passes through the small pipe bypass and into the meter. When a suppression system is activated and there is a fire flow in the main, the demand flow of the suppression system raises the clapper, allowing a full flow through the check valve. With this detector check valve, only small flows of water are metered; fire flows are not. The concept is that with a suppression system, the fire would be controlled and extinguished in a relatively short time, with the fire department using unmetered public water from the fire hydrants just for mop up. The detector check valve may also be used in the supply line to serve as a regular check valve and must be UL listed and/or FM approved.

Key Valve

A key valve is a non-rising stem valve installed in the underground piping. Covering the operating nut on the top of the valve is a round metal shaft known as a roadway box that fits over the top of the valve and extends to the surface of the ground or pavement. A metal cap fits on the top of the roadway box and cast in the metal of the cap is the word water. To operate this valve, the cap is removed, and a long-handled metal wrench is slipped into the roadway box and fitted over the valve operating shaft. The operating wrench, or key, has a crossbar to facilitate turning and opening or closing the valve. Since the key valve is not an indicating valve, it is never to be used as a control valve on water supplies to suppression systems.

Fire hydrants require periodic maintenance. Those on private property are supplied from the fire main that feeds the suppression systems. Key valves are installed on the supply to the fire hydrant so that when the hydrant requires mainte-nance, the key valve can be closed, eliminating the necessity of shutting down the entire fire main. Hydrants are important firefighting equipment, and it is much more practical to have one hydrant out of service than the whole system shut down for repair. All valves must have the UL listing and/or FM approval.

Full Flow Fire Meters

When a meter is necessary for both small and full flows in the fire line, a full flow fire meter must be installed. These meters are large and expensive compared with the detector check meter and are installed between two OS&Y valves. Full flow fire meters may measure approximately 45″ for a 6″ line, and 53″ for an 8″ line. The code authority having jurisdiction may also require a bypass around the full flow meter. Should the meter be out of service, the normally closed bypass would be opened. When a bypass line is required, the overall measurement of the meter and bypass connections increases 98″ for a 6″ line, and 112″ for an 8″ line-face to face of the flanged bypass tees. Full flow meters are required by city ordinances, when the fire line is used to fill a reservoir or when testing deluge systems. The requirements of the local code authorities will determine their needs for metering the fire water lines. Most municipalities and authorities do not require metering, but if this is not determined up front and is later discovered, it can be a costly addition to the system, especially if full flow metering is required.

Backflow Preventers

Fire protection water in sprinkler systems may lay dormant for years and, as a result, can become contaminated. A pendent head that has been installed for a number of years, untouched, contains water trapped in the drop. This contaminated water might find its way back into the potable drinking water if the sprinkler system is connected to a public water main and a flow reversal occurs. Backflow preventers, installed between the potable water supply and

Fig. 7.3 Backflow
Preventer with Open
OS&Y valves (Robert
Till)

the sprinkler system to prevent such contamination, are not only becoming more popular, but have become mandatory in most states and cities.

There are two basic types of backflow preventers: the double check valve arrangement (Fig. 7.3) and the reduced pressure type. The installer is not given a choice of which type to use; state or local requirements specify the type. Both types must always be installed between two shut off valves, preferably OS&Y valves. The double check valve backflow preventer does not use standard check valves. They are spring loaded check valves, and the entire arrangement is UL listed. The double check valves in this type of backflow preventer can be detector check type with a small metered backflow line around the preventer. Some models of backflow preventers can be installed in the vertical position, and others horizontal only. The backflow preventer should be installed immediately ahead of any system risers where the underground supply line enters the building. The reduced pressure backflow preventers require provisions to facilitate the use of test connections. Backflow preventers must be UL listed and/or FM approved.

Firefighter Intervention: Manual Fire Suppression

8

The Fire Service: Introduction

Fire department tactics are usually a function of the structure being defended. Because of the large lumber industry in the US, there is a tendency to construct structures of wood. The configuration of wood in these structures leads to fires that may not only consist of burning the contents of a structure, but also the structure itself. Many countries outside the US, depending on their location and legislation, may use more brick and stone in their construction. This leads to more contents fire and less structural fires, and therefore sometimes completely different fire department tactics. The "manual suppression coordinated operations" very generally discussed here are those of the US.

Architectural and site conditions for a building can dramatically affect an initial fire attack and the tactics used by the fire service to extinguish a fire. When architectural obstacles and difficult site conditions delay the initial fire attack, the fire grows and propagates causing additional property damage and greater danger to occupants and firefighters. Callery describes the interactive nature of the process as follows:

> When a fire department quickly extinguishes a challenging building fire, the firefighters usually take the credit for making a "good stop". Often the building deserves much of the credit. Likewise, when a fire gets away from the department, the building deserves some of the blame. The likelihood of quick extinguishment will differ for various fires of similar size at the time the first fire apparatus arrives. The change in impact is brought about by the building and the location of the fire within the building (Callery 1996).
> Captain Jim Callery – Worcester, Massachusetts USA Fire Department

Established Burning to Arrival

The fire department time line, Fig. 1.1 in Chapter 1, begins with established burning. The line then sequentially recognizes that detection, fire department notification and arrival are requisites to the time duration needed to apply to first water. The time from established burning to fire department notification is generally a function of building operations in the structure of origin as shown in the figure.

When modeling the time to suppress a fire after fire department notification, it is necessary to acquire data on individual tasks to develop the model and whole evolutions to validate the model. The time to fire department arrival can be estimated from statistical data of the local community. A variety of different techniques may be used for this analysis. For example, simple speed and distance estimates may be used.

One direct solution to the time between fire department notification and arrival first water is

R. C. Till, J. W. Coon, *Fire Protection*, https://doi.org/10.1007/978-3-319-90844-1_8

Fig. 8.1 Fire timeline

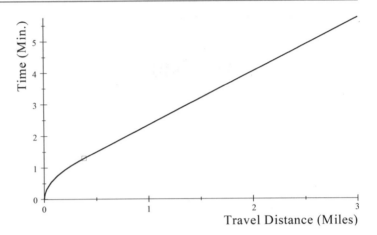

shown in Fig. 8.1. In this study it was found the following equations provide travel time estimates.

Figure 8.1, is derived from the equations below. These equations are based on analysis of data from Trenton N.J., Denver CO, Wilmington DE, and Yonkers, NY. Rand Study of 1979 (Warren et al. 1979).

$$E(D) = \begin{cases} 2.10\sqrt{D} & \text{if } D < .38 \\ 0.65 + 1.70D & \text{if } D \geq .38 \end{cases}$$

It can be seen that a travel distance of two miles requires about 4 min. for fire department arrival according to this model. The study found that this correlation worked well for most cities. It should be noted that the time from notification to arrival must also include alarm handling and turnout times. These time durations, combined with the transit time, provide the total response duration.

Arrival and Size Up

The first arrival is usually an engine company. This first-in company places itself in what appears at the time to be the most appropriate position to carry out initial activities. This positioning involves balancing several immediate and possibly conflicting needs including gathering information about the location and magnitude of the fire, ascertaining if there are any potential life safety threats, establishing a water supply, and laying attack lines. Shortly after the first company arrives, a chief and the other companies may arrive.

The fire ground commander is the senior officer at the fire scene. Initially, it is the officer of the first arriving company. However, this position is transferred when a more senior officer arrives. The fire ground commander is in responsible charge of the scene and all of the operations. The fire ground commander is placed in a position of making rapid decisions on the basis of incomplete, often inaccurate, and constantly changing information. In addition, these decisions must be made during periods of maximum distraction and emotional stress within very short periods of time. The process becomes decision making under uncertainty in a very pure form. Experience, knowledge, and training are very important to a fire ground commander's educational background. Timely, astute decisions can make the difference between a good stop and a fire that got away.

The other important element to the success or failure in limiting the fire to acceptable sizes is the influence of the building site and design features. An initial size up becomes the basis for the fire ground commander's decisions. A size up is a mental model of the entire system that makes up the fire ground conditions at the time of arrival. A size up begins on route and becomes more defined upon arrival at the site when better information becomes available.

The goal of fire ground operations is to use the available community resources of manpower and equipment to rescue endangered individuals and to minimize the destructive damage of a hostile building fire. The size up provides the information on which to base decisions regarding the strategy and tactics to achieve that goal. An ideal size up would include the type of information described below. Obviously, time constraints of required early action preclude the collection of all of the information initially. However, when complete information is not available, the fire ground commander makes decisions based on the accumulated knowledge at the time.

- Life Safety – Are any occupants remaining in the building? Where might they be located? How certain is the information?
- Building Construction Information – In buildings of this type, will the barrier effectiveness quality contain a fire to one or few rooms or will it allow the fire to propagate easily? What is the collapse potential? Can the building construction be used to advantage to control the fire and limit its propagation? Does this building present a safety hazard to fire fighters
- Fire. – Where is the fire? What is its size? What is burning, and what is the fuel loading? What types of hazardous materials may be present? Are they involved in the existing fire? Can they become involved if the fire extends?
- Water Supply – How many hydrants are available, and where are they located? How much fire flow is available? Are static water supplies available for drafting? Must water be supplied by tankers?
- Fire Fighting Resources. – How many engines and fire fighters are available? What other types of apparatus are present or available? Where is the equipment now placed? Should they be relocated? If additional equipment will respond, where should they be placed? Are personnel needed for a primary search and rescue or are they available for firefighting? How many hose lines could we lay? What fire flow can be delivered internally and externally?

- Fire Attack. – Where are the building access locations? Are there any natural or man-made obstacles to approaching the building from different sides? Where are the stairwells? Where are the standpipes? How do we gain access to the floor of origin? What fire attack routes are available? How clearly are they recognizable? Will doors have to be forced to gain access to the fire location? How many? What size and how many hose lines can be stretched with the manpower available? How long will it take to lay the hose lines? Are there critical locations at which to position hose lines in order to defend people or valuable property from fire extension? Will these positions help or hinder fire attack or control? Where and how can the building be ventilated.
- Environmental Conditions – What is the time of day? What are the weather conditions? What are the site conditions that influence movement, such as mud or snow? How will the wind conditions affect firefighting? What are the smoke, heat, and visibility conditions within the building? What are the heat and visibility conditions near the fire?
- Exposures – Are there any other external buildings exposed to the fire? Should those exposures be protected before or subsequent to attacking the existing fire? Are there internal exposures that should be protected?
- Building Services Information – What type of heating fuels are used, where are they stored, and where is the fuel shut off located? What is the electrical system and where is the shut off? Is emergency lighting available? Does the HVAC system continue operation, shut off, or shift to an emergency mode of operation at detection? Can its operation be changed to help firefighting? Where are the controls and how is it done? Are elevators in operation or is there a recall at detection? Is there a fireman's key for emergency use? What other building services information is available that would influence life safety or firefighting?
- Resource Augmentation. – Does standpipe water need to be augmented from the build-

ing's fire department connection? Does the building have a sprinkler system at the fire location? Would connection to the sprinkler Siamese enable the system to operate? Are additional alarms or mutual aid needed?

The more information that is available, the better the decisions can be made. However, as the fire continues to burn, the conditions change. The fire companies will take initial actions in accordance with a preconceived plan of operations and immediate needs before a size up is completed. This plan will be modified by the fire ground commander to reflect conditions that are evident or that change during the course of the fire. Communications and the chain of command become essential features in fire ground operations. In many instances, communication becomes the basis for the transmission of new information about conditions inside the building to the command post.

Communication is also crucial to the proper deployment of resources and to learn of the outcomes of the actions taken (Till 2001).

Manual Suppression Coordinated Operations

While life safety may be the principal concern for the fire ground commander, the focus of this chapter is on building evaluation for its ability to limit fire extension by manual suppression. Manual suppression involves five coordinated operations. They are,

1. find the fire
2. establish a water supply lay attack lines and
3. initiate agent application to the fire
4. prevent extension of the fire to other spaces
5. extinguish the fire.

Depending upon the size up, the fire ground commander may employ an interior offensive attack to extinguish the fire aggressively. Alternatively, the commander may select an exterior or interior defensive procedure to pre-

vent fire extension to threatened exposures outside or within the building.

Building and fire conditions may change over time, requiring the fire ground commander to alter tactics from an offensive mode to a defensive mode.

Locating the Fire

Although locating the fire may seem to be a trivial problem, the process is not as easy as it at first may appear. Of course, small buildings do not usually pose much difficulty. However, in large or complicated buildings, finding the fire can be time consuming and difficult. Dense smoke obscures visibility, and a fire often is located by feeling the heat rather than by seeing the flames. Consequently, accurate information given to the fire officer can substantially reduce the time to agent application in a complicated building. Accurate information on the fire location is important not only to place the attack hose lines effectively, but also to become aware of potential life safety or property protection concerns.

Establish a Continuous Water Supply

Another part of the process that must be evaluated is the establishment of water to supply the attack lines. Engines will carry some water in their tanks. Often, the 500 or 750 gallons is sufficient to provide enough water to extinguish a smaller fire or to start a fire attack, before the fire department connects to hydrants or provides other means for a continuous water supply.

Standpipes: Minimizing Set Up Time

Standpipe systems are piping systems that provide fire hose stations for manual application of water to fires in buildings. Standpipes are classified in three classes. For more advanced coverage of this topic, the reader is referred to Isman (Isman 2016).

Class I Standpipe

This category of standpipe supplies 2–1/2″ hose outlets at each floor level, usually located in stairtowers, for use by firefighters or those (thoroughly) trained in handling heavy 2–1/2″ hose fire streams. These 2–1/2″ outlet hose valves furnish the firefighters with a hydrant at each floor level, and an adequate water supply to provide an effective hose stream for use during the advanced stages of a fire. A live hose outlet at each floor level installed in a fire-resistive stairtower to protect the fireman ascending to the fire floor allows the fireman to carry dry hose up the stairs in lieu of dragging a connected hose up the stairs. The fireman can control the flow of water from the floor below the fire floor to reduce the possibility of fire involvement in the stairtower at the point of hose connection when the stairtower door is opened on the fire floor.

Water Supply

Class I standpipes must have an adequate water supply to furnish sufficient volume and pressure for 2–1/2″ hose streams. The water supply for Class I standpipes must provide a minimum of 500 gpm if only one standpipe is required, and 250 gpm for each additional standpipe. For example, if a building requires four standpipes, the watersupply must be capable of providing 1250 gpm. The total supply demand shall not exceed 2500 gpm, and based on the 500 gpm for the first standpipe and 250 gpm for each succeeding standpipe, a maximum of nine standpipes.

Pressure Requirements

The pressure requirement for Class I, and Class III standpipes is a water supply that will provide 100 psi residual pressure with the required gpm flowing at the highest outlet on the standpipe.

The pressure requirement for Class II, standpipes is a water supply that will provide 65 psi residual pressure with the required gpm flowing at the highest outlet on the standpipe. Residual pressure is the pressure reading on the gauge with the required gpm flowing. (Residual pressure will be discussed in Chap. 10.)

A method commonly used in high rises to maintain operating pressure is to employ pressure reducing valves (PRV's). However they need to be set properly, as demonstrated by the One Meridian Plaza fire (Philadelphia, PA) February 23, 1991.

Class II Standpipe

A Class II standpipe system is one designed to be used by the occupants of the building until the arrival of the fire department. It is sometimes referred to as a small hose standpipe because it supplies 1–1/2″ hose stations. Note hose stations, not hose valves, are supplied by a Class II standpipe system. It is extremely dangerous to have 2–1/2″ hose station with hose connected to the 2–1/2″ hose valves on a Class I standpipe. If untrained individuals attempt to use the hose to fight a fire, they could be seriously injured from the excessive thrust produced by the flow of water under pressure. Handling 2–1/2″ hose streams is a job for trained firefighters; not a one man job.

Some city building codes require 2–1/2″ hose at each outlet to eliminate the need for firefighters to carry the hose up the stairs. Although a good concept from the firefighters standpoint, there should be some latitude for judgment because there is always a good likelihood that an occupant could attempt to fight a fire with the 2–1/2″ hose. The possibility of serious personal injury and unnecessary heavy water damage may outweigh the benefit to the firefighters.

There is another important reason for not furnishing 2–1/2″ hose. The 2 1/2″ hose outlets on the standpipe are usually (and should be) located in a stairtower that is constructed to provide safe exit for occupants and a safe access to the fire floor for fire fighters. If an occupant connects the 2–1/2″ hose to the hose valve and drags the hose through the stairtower door to attempt to fight a fire in the building, the hose will prop open the stairtower door allowing smoke and heat to enter the stairtower which could make the stairtower impassable for occupants attempting to exit the building.

Even a small amount of smoke and heat in an exit stairtower can cause disorientation of building occupants, which can contribute to death and injury as easily as fire. People coming down the stairs in the stairtower from the upper floors may trip over the extended hose in their haste to escape.

One more consideration against placing a supply of 2–1/2″ hose by the 2 1/2″ hose valve is the likelihood of an untrained occupant dragging the hose to the fire area in the building and, when he opens the nozzle, the hose is whipped from his hands. In the confusion of a live hose crashing about, the novice firefighter will run for the exit. By the time the fire department arrives, they are faced with a fully involved fire on the floor of fire origin, a partially opened stairtower door on the fire floor making access to the hose valve at that floor level impossible, and a 2–1/2″ hose wasting water and reducing water supply volume and pressure needed to combat the fire.

Water Supply

The water supply for Class II standpipe service must be capable of supplying 100 gpm at the highest outlet on the standpipe with a residual pressure of 65 psi with the 100 gpm flowing. A Class II standpipe supplies the 1–1/2″ hose rack or hose rack in a cabinet, that are commonly found in commercial, industrial, and institutional facilities. These hose stations provide a means of controlling a fire in the early stage when used by the occupants of the building.

Pressure Requirements

One hundred psi is the maximum pressure allowed at any hose valve outlet on a standpipe system. This maximum pressure requirement is especially important to safeguard the untrained occupant using a 1–1/2″ hose stream for the first time. Even with a 1–1/2″ hose, high pressure discharge in inexperienced hands can make control extremely difficult and dangerous. Standpipes that have a water supply that provides a standpipe pressure in excess of 100 psi often occurs when a standpipe serves a multi-story building or is supplied from a fire pump. To ensure a pressure of 65 psi at the top-most outlet of 100 high building means that a pressure of at least 143 psi must be provided at the first floor. Disregarding friction loss to transport water to the 100′ hose outlet and merely adding the static pressure of 0.434 psi per foot which must be overcome to lift the water up to the 100′ elevation. Pressure and flow will be fully explained in Chap. 10.

To reduce the various pressures at the hose outlets on different floors to 100 psi or below, special hose valve equipment must be installed. The simplest method is to install a pressure-reducing disc in the hose valve outlet. This disc consists of a brass plate with an orifice of a predetermined size drilled in the center. The pressure reducing disc is quite efficient if the standpipe pressure is relatively constant because the orifice size can be determined by a calculation of the standpipe pressure at the specific hose outlet.

Current state-of-the-art practice is to use a hose valve that is not factory set for a specific pressure reduction, but that automatically adjusts the inlet pressure to a safe outlet hose pressure.

The Class II hose stations should not be located in an exit stairtower: The obstruction the hose would present to people using the stair to exit the building, and the propped – open stairtower door on the fire floor. The hose valve referred to can be either an angle valve or straight-way valve connected to piping from the standpipe riser. This valve can be opened before the hose is removed from the hose rack without emitting water from the nozzle. The folds in the hose prevent water from flowing until the last fold comes off the rack. Water will then flow from the nozzle unless it has a device to control the flow. Some nozzles not only have an on-off control, but also a means of changing the flow from a straight stream to a water-fog, and some of these combination nozzles have a method of adjusting the water-fog pattern. The selection of a nozzle for 1-1/2″ hose is extremely important and must be based on a thorough evaluation of the hazard and user.

Improper application of a straight stream hose discharge has the potential to spread the combustion by scattering the burning material. For example, when directing the stream from a garden hose into a pocket of mud, the mud is scattered in all directions. The same situation will occur when an 1 1/2″ straight stream is directed onto burning combustibles like paper, trash, etc. For example:

An accidental flammable liquid spill that was ignited should never be fought with a straight stream water discharge. Like the mud puddle, the burning liquid will be scattered over a wide area. With this fuel spill fire, a combination nozzle, and an individual with some training in the use of small hose firefighting, the fuel fire can be controlled with the water spray nozzle adjustment. In addition, the straight stream nozzle adjustment can provide a discharge onto material and equipment that is exposed to the radiant heat of the combustion to maintain a temperature that will prevent heat damage and/or prevent flammable liquid drums and tanks from exploding.

Some nozzles discharge only a very fine spray pattern of tiny water droplets. This water pattern is like an extremely heavy fog. The spray pattern absorbs so much combustion heat that a person can advance close to even an intense fire without being overcome by the heat. This fog pattern will not scatter the burning material, and the water droplets in changing to steam absorb sufficient heat to control and extinguish the combustion. Another important consideration is the presence of live electrical equipment. A solid straight stream discharging onto electrical equipment, to especially high voltage equipment, can act like a wire and conduct the current back to the nozzle and the user. Unlike the straight stream, discharge of the spray pattern on electrical equipment does not have the

potential to conduct an electrical current. However, the voltage of the electric equipment and the closeness of the nozzle to the equipment could produce an electrical shock to the hose operator.

The hose rack on which the hose is hung is usually a semi-automatic type. The hose is hung on pins, and it is released as the nozzle is pulled away from the rack.

1–1/2″ Hose Stations

Small fires can be extinguished with a small hose before they grow into a major conflagration, or can be controlled until the fire department arrives. Hose stations with 1–1/2″ hose for occupant use may be a curse or a blessing. Time is the greatest ally of fire, and unless there is a water flow alarm on a Class II standpipe that will automatically alert the fire department and activate local alarms, there is always the possibility that an occupant will not call the fire department right away. In an attempt to fight a fire, they will not call until realizing it is beyond their ability to so do, and then summon the fire department and alert other occupants. Meanwhile, the fire has ample time to develop before the firefighters arrive or the alarm is sounded.

These advantages and disadvantages must be given serious consideration when a standpipe system is being evaluated.

Class III Standpipe

The third standpipe system is called a combination standpipe and is classified as a Class III standpipe. The Class III standpipe system is a combination of a Class I and Class II standpipe. The Class III standpipe supplies both 2–1/2″ hose outlets at each floor level in the stairtower, and Class II 1–1/2″ hose stations located throughout the building outside of the stairtower.

On Class I and Class III standpipes it is sometimes required, and always a good idea, to install a 2–1/2″ by 1–1/2″ removable reducing coupling in the outlet of the 2–1/2″ fire department valve. With this coupling, the fire department can connect their 1–1/2″ hose or remove the coupling and connect their 2–1/2″ hose. The coupling is usually a pin lug type with two pins that fit the hole on a spanner wrench for quick removal of the coupling.

Standpipe Water Supply

The water supply for standpipes can be furnished by a connection to a city water main, through a fire pump to boost the pressure, gravity tank, or pressure tank. If the supply is from a limited source – such as a tank in lieu of an unlimited supply from a city water main-it must be of adequate capacity to furnish the total demand for at least 30 min.

Types of Standpipes

There are two types of standpipe systems: a wet standpipe and a dry standpipe. Like wet sprinkler systems, a wet standpipe pipe is constantly full of water under pressure and is used when the area is heated to prevent freezing.

There are two types of dry standpipe systems. One type is dry because it is not connected to any permanent water supply. A standpipe with fire department 2–1/2″ hose valves at each floor level provides firefighters with an outlet when the fire department pumper truck pumps water into the standpipe riser(s) through a pumper connection similar to the siamese connection on a sprinkler system.

The total demand for a standpipe being used by the fire department may be more than a standard 2–1/2″ × 2–1/2″ × 4 siamese connection can provide. Depending on the size and number of standpipes, the fire department pumper connection on either a wet or dry standpipe may have four 2–1/2″ outlets manifolded into a 6″ supply pipe.

The dry standpipe without a permanent water supply is used in unheated areas, areas where the water supply is questionable, and buildings under construction where the pipe is exposed and could freeze or where the water supply is not completed during construction. Dry standpipes used during construction may remain dry or be converted to a standard wet standpipe when the construction is enclosed, heated, and a water supply connection complete.

When a dry, no water supply, standpipe is installed, and rises several floors above the pumper connection, a check valve should be installed in the main supply line on the standpipe system side of the pumper connection. With the riser(s) full of water under a static head, or elevation pressure, the small clappers in the siamese connection cannot be expected to retain the water under pressure. When hoses are removed from the siamese, a backflow can result. The check valve can hold water in the riser(s) until it can be safely and conveniently drained off through a drain valve installed on the standpipe system side of the check valve.

An important feature which must be incorporated into wet standpipe design for all three standpipe classes is a main control valve at the base of each standpipe riser, on the water supply side of all hose outlets on each riser. This valve is used to control the water supply to each riser in case of a break in the standpipe or if a hose that is discharging has been abandoned because of some unexpected firefighting condition, (as in the example of the occupant using a 2–1/2″ hose). A control valve on each riser gives the fire department control of each standpipe riser if, for any reason, water volume and pressure is being wasted. Another reason for this individual riser control valve is that should a riser be down for repairs, additions, or alterations, only that one riser will be shut off and the others will remain in service and ready for immediate use.

The second type of dry standpipe is also used in unheated buildings or stairtowers and is sometimes referred to as a remote control fire hose standpipe. Except under unusual circumstances, the dry standpipe with no permanent water supply should only be a Class I standpipe.

The remote control dry standpipe system design is similar to the wet Class I, Class II, or Class III standpipe, but the piping is empty. The water is held back by a deluge valve installed in a warm area of the building or a heated valve house.

A remote control fire hose station, like a break glass manual fire alarm station, is installed immediately adjacent to each hose station. Each of these remote-control stations is connected to the release mechanism of the deluge valve, and water fills the system when the manual station is activated. Some remote-control systems electrically activate the deluge valve release mechanism, but to eliminate the possibility of power failure making the standpipe system inoperative, several manufacturers have developed pneumatic systems. Operation of a remote manual station releases compressed air which, in turn, trips the deluge valve and water fills the standpipe system.

One dry standpipe system consists of remote control stations that function as follows: When the small glass window on the face of the remote manual box is broken with a small hammer (which is attached to the unit by a chain), a spring is released which operates a bellows sending an air pressure impulse to the release mechanism of the deluge valve.

If air tight hose valves are used on a dry standpipe system, it is possible to use a regular dry pipe valve to control the water supply with the piping filled with air under pressure. When a hose valve is opened, compressed air is released, and the dry pipe valve releases water to the system. The remote station deluge valve system and the dry pipe valve system standpipes introduce an element that is not encountered with a wet pipe standpipe. Once the dry system is activated by operation of a manual remote station or opening a hose valve on the dry pipe valve system, there is a delay in water reaching the nozzle. This charging period can produce some unusual hose action when it is suddenly pressurized with water. A test conducted on a dry standpipe using remote stations (the spring and bellows) indicated that 11 s after activating the remote station 333' from a 6" deluge valve in the basement of the building, water was discharging from a 2–1/2 hose line. It is obvious that water rushing through piping and into hose at a velocity of 30 per second or more can create hose and nozzle reaction that could be dangerous, especially in inexperienced hands. If an individual is knowledgeable of the operation of the system, the remote station can be activated

leaving the hose valve closed until the system fills with water and then opening the hose valve and removing the hose. Once the water reaches the closed hose valve, the system is a wet pipe standpipe system.

Hose Station Distribution

The distribution of hose stations is a matter of engineering judgment. Hose stations should be located so that each section of the protected area, and all portions of each floor, are within 30' of a nozzle attached to 100' of hose. Hose for use by occupants longer than 100' becomes unwieldy, could lead to twisted or kinked hose limiting the flow, given the fact that all the hose must be pulled off the rack before water will flow. This location of 1–1/2" hose stations is a minimum requirement, but the number of standpipes, location, and distribution of hose stations depends on occupancy, construction of the building, accessibility of hose outlets, and number of types of hazards. The standpipe risers with 2–1/2" hose outlets should be located where they will be protected from mechanical and fire damage and exposure, and readily accessible in a fire condition with the center line of the hose outlet approximately 2'–6" above the floor.

Each occupant-use hose station should consist of a 1–1/2" hose valve supplied from the riser. An automatic drip drain must be installed at the hose valve when linen hose is used to prevent any leakage past the hose valve from entering the hose. The hose nozzle may be a straightway pattern with no shut off valve and a 3/8" orifice. When the hose is pulled off the rack with the hose valve, open water will discharge. The hose should be rubber or Neoprene lined, as lined hose greatly reduces the friction loss, with an outside woven jacket. Unlined linen hose has been used quite extensively because it folds easier and fits into a smaller space, and initial cost is less than lined hose. Unlined hose will mildew and rot if left damp or if stored in a damp atmosphere, and unless dried in a professional manner it is, in reality, a one-shot use hose. Firefighters usually do not know if the hose stations are stocked with unlined hose that may crumble when pulled off the rack, or even lined hose that may be old and in an unreliable condition. In most cases they will use their own 1–1/2″ hose.

If unlined linen hose is stored in a dry, heated atmosphere, it should last indefinitely, and for emergency use it is the most economical hose. On a construction project to serve the hose outlets of a standpipe that is installed as the building construction progresses, and with no intent to use the unlined linen hose in the finished building hose cabinets, unlined linen hose will serve an excellent, economical, purpose. This hose is usually constructed of 100% linen flax yarn to meet Underwriters Laboratory (UL) listing and Factory Mutual (FM) approval requirements, and it can be given a mildew treatment to retard deterioration.

Most lined hose has a cover of cotton wrap with a Dacron or polyester filler, either single or double jacket. The lightweight polyester single jacket, neoprene-lined (tube) can be used on pin racks, and occupies only a little more room on the rack than unlined linen hose. Some hose construction like single jacket and double jacket consisting of cotton/polyester with a neoprene tube lining cannot be folded to go on a pin rack and must be stored on a hose reel or hump rack. A hump rack is a rack where the hose is piled in horizontally, one fold on top of the next. For many years, the size of the standpipe was governed by the height of the standpipe as follows:

(A) Standpipe not exceeding 75′ in height or 100′ in length, 4″ in size.

(B) Standpipe in excess of 75′ or 100′ length, 6″ in size.

(C) Standpipe height limit 275′ and when a building is in excess of 275′, the standpipe system will be zoned. (Standpipe zoning in highrise buildings is explained at the end of this chapter.)

Note that the sizing requirements use the phrase, in height or length. A piping system supplying hose stations throughout a one-story building is also referred to as a standpipe system.

With the advent of extensive use of hydraulic calculations for sprinkler systems in lieu of the pipe schedule sizing, it is recommended that standpipe systems be afforded the same engineering status. By hydraulically calculating a standpipe system, the pipe sizes can usually be greatly reduced, but of more importance than this economic advantage is the fact that calculations will determine that the flow and pressure at each hose outlet will be adequate to meet requirements.

Hose racks in buildings where appearance is a factor, are usually stored in cabinets that are completely recessed in the wall with the door flush with the wall, or semi-recessed with 1–3″ extending beyond the wall surface. Surface mounted cabinets are not recessed into the wall and are mounted on the surface of the wall or in a column.

When designing a standpipe system and locating hose cabinets, it may be necessary to recess a cabinet in a wall that is rated to resist fire spread. Fire walls are constructed to prevent fire from spreading from one area to another for a specified time. These rated walls enclose stairtowers to protect people from the heat of a fire and smoke while they are exiting the building. Recessing a cabinet in a fire wall in effect creates a weak point in the fire wall. Should the thin metal cabinet be distorted by the heat of the fire, the cabinet can drop out of the fire wall and allow smoke and heat to pass through, involving personnel and material on the protected side of the fire wall.

If it becomes necessary to locate a recessed cabinet in a fire wall, the wall can be extended around the back, top, and bottom of the cabinet. If the cabinet should be lost, the integrity of the fire wall would not be destroyed.

The center line of a 1–1/2″ hose valve should be located 5′–6″ to 6′ above the finished floor, for ease of operation. Unless local conditions dictate a different arrangement, the hose valve is usually an angle valve with a nipple, called a rack nipple, which screws into the valve outlet and supports the hose rack. The hose coupling connects to the rack nipple. The hose rack swings on the rack nipple a full 90°, and this operating distance, combined with the cabinet door swing of 170° (to give full and unobstructed access to the hose rack in the cabinet), must be considered in designing the location of racks and cabinets. Some hose cabinets are sized to accommodate both the 1–1/2″ hose and a separate 2–1/2″ hose valve. Also available are cabinets for individual 2–1/2″ hose valves.

Roof Hose Stations

Some building codes require hose outlets on the roof of a building to provide firefighters with a hose outlet to fight roof fires ignited by embers from an exposed fire, a fire in an elevator, or mechanical penthouse, or to direct water to a fire in an adjoining building. These roof outlets-known as roof manifolds-consist of two or three 2–1/2″ hose valves with caps over the openings to prevent foreign matter from falling into the valve. The supply for these roof manifolds is a 4″, 6″, and 8″ extension above the roof of the standpipe riser. Since wet standpipes and the hose manifolds are exposed to the weather and subject to freezing, an indicating control valve is installed at hand height in a heated area at the floor level below the roof and opened when the roof manifold is used. An automatic drain is installed on the roof monitor side of the control valve to make sure the pipe and roof monitor remain dry when the control valve is used.

The roof manifolds are 2–1/2″ hose valves because they are hydrants on the roof for use by the firefighters. If there is a possibility of fires at the roof elevation that could be controlled or extinguished by building occupants or a company fire brigade, roof hose storage cabinets are available for 1–1/2″ hose. These roof hose stations are supplied from piping on the ceiling of the floor below, but instead of controlling the supply to the roof hose station from the floor below the roof, the valve is installed in the piping on the ceiling below the roof. The handle of the valve extends through the roof and into the roof hose cabinet. The drain valve is installed like the handle, also extended into the hose cabinet. The cabinets contain 100′ of 1–1/2″ lined, hose and nozzle, with

the hose folded horizontally under a waterproof cover that unlatches and lifts up on hinges.

Standpipe Zoning

The maximum height for standpipe systems is 275′, and when a standpipe must exceed this height, the system must be zoned. A zone signifies the upper and lower levels of a building served by standpipes. When zoning a standpipe, 275′ is maximum height, but the zones may be less than 275′.

An example would be a building 355′ in height. The lower level, or zone, could be 275′, with the upper level of 80′, but a more equal zoning would be a lower zone of 200′ and an upper zone of 155′. If the two zones had exceeded 550′, the building standpipe system would have been extended to three zones. In the two-zone example, the lower zone would have been a standard standpipe arrangement, with the possibility of a gravity tank at the top of the zone supplying the lower zone standpipes.

The upper zone standpipes would be fed by express risers or risers that do not supply water to any system below the base of the upper zone standpipes. The upper zone standpipes may also have the additional supply of a gravity tank. Due to the excessive height of both of these zones, a fire pump will undoubtedly be required to provide sufficient pressure to overcome the static (elevation) pressure loss and still furnish 65 pounds at the topmost standpipe outlet. In the example, the topmost outlet of the upper level is 355′ above the supply, and it would take 154 psi just to raise the water to this height. This is one reason for the gravity tank above the standpipe zone elevation to maintain positive pressure on the system.

Each zone of high rise zoned standpipe systems requires a separate siamese connection. According to conditions, one fire pump may supply the lower zone in the example and the lower zone gravity tank, and the second fire pump located at the bottom of the upper zone takes its suction from the lower zone gravity tank. This arrangement prevents the upper zone from having to overcome an additional 87 psi static pressure to get the water to this elevation. For multiple zones, additional pumps for the upper zones will be required. These pumps are located at the base of the higher zones. The lower pumps supply water to the gravity tanks, and the upper zone pumps take suction from the gravity tanks.

Combined Standpipe and Sprinkler Riser

Regulations allow a single riser pipe to supply both sprinklers, hose stations, and 2–1/2″ hose outlets. Additional equipment is required for each floors sprinkler supply coming from the riser such as a control valve, drain, and test connection. The combination sprinkler and standpipe riser installed in the stairtower supplies the 2–1/2″ fire department valve at each floor level. If 1–1/2″ hose stations are required – although they are usually not required when the building is totally sprinklered – the supply for these hose stations can be piped from the riser throughout the area.

Water Supply

The supply for sprinklers is obtained from the riser at each floor. This supply to sprinklers at each floor must be controlled by an indicating type control valve.

Should the sprinkler system on one floor of the building have to be repaired or remodeled, it will only be necessary to shut down this one floor system, leaving the balance of the sprinkler and standpipe systems in service. With the water supply valve shut, but the piping full of water, it is necessary to provide a drain valve on the system side of the floor control valve. This drain valve must be connected to a drain line to discharge the system water to a safe location. In most high-rise buildings, a sprinkler water flow alarm is necessary at each floor so the fire department can locate the floor where the fire

has activated sprinkler heads. Each sprinkler supply from the riser at each floor is provided with a paddle-type flow switch on the system side of each floor control valve. It is also necessary to test the individual flow switches and, in addition to the drain, a test valve must be provided. The combination drain and test line with drain and test valves can be piped as follows: A restricted orifice giving the flow access to one sprinkler head to operate the flow switch must be installed in the test line. There are devices available that combine the drain and test in one single element, and these can minimize the space required to individually pipe the drain and test.

Standpipe systems provide adequate equipment to combat fires in different stages of growth. The small hose stations give occupants the maximum opportunity to provide help in a fire emergency, and 2–1/2″ outlets provide firefighters with effective equipment to combat a fire that has reached major proportions. Standpipe systems must be designed based on a complete study and evaluation of all facets of the structure, occupancy, hazards, and personnel. Standpipe systems are not a substitute for sprinkler systems, but they can be a tremendous complement to automatic protection systems.

Set Up Time: Firefighting Tasks

Research identified tasks that needed to be completed to set up a water supply from a fire hydrant – all shown to correspond not to Normal (bell shaped) distributions, but to skewed, lognormal distributions, as show in Fig. 8.2 below. This is not surprising since tasks involving humans tend to take this form (Till 2001). It should be noted that since the tasks are skewed it is very likely that the probability of overall attack time will be skewed as well, and that summing up bell shaped distributions will not in all likelihood produce a valid response distribution.

A method of summing these times is known as Discrete Event Simulation (DES). This method allows the simulation of water supply setup tens of thousands of times, even millions of times, and therefore the development of overall curves for a setup for a particular fire. In time this could be adapted for internal attack, but the data is not yet available. The use of helmet mounted cameras (Gopro etc.), and the understanding that this is NOT a competition will make this data easier to collect over time.

Set Up Time: Firefighter Factors

Many factors will influence the time it takes for firefighters to set up water supplies and apply attack water. These include:

Low water pressure	security	ignorance of condition
Unfamiliar neighborhood	no windows	arson
icy weather	remote location	poor construction
false alarm	no water	not enough men
bad equipment	dangerous chemicals	firefighter deaths
obsolete equipment	potential for explosion	firefighter injury
unfamiliar area	poor training	unfamiliar command

Changes in any of these factors can change the overall response to a fire.

Set Up Time: Location of Fire Within a Structure

If the presence of fire is not known, it cannot be responded to. Detection of fire may be non-trivial for a variety of reasons, but in most situations, it means that heat or the products of combustion (smoke) are discovered. It may be non-trivial because our sensor (detector) is blocked from the fire, or that the products of combustion have characteristics that are difficult to detect. Both of these situations will lead to longer detection times.

Interior Attack

A routine procedure can be developed to determine the time to lay fire attack lines along a specified route within a structure and to coordinate that operation with supply water establishment. The creation of such an analytical method that incorporates fire service experience to predict the fire size at the time of initial water application will provide such benefits as,

- numerical measures that allow local firefighting resources to compare building designs.
- fire department development of site specific operational planning for available local resources and conditions.
- comparison of design alternatives to give the local fire authorities additional ways to communicate with owners, architects and other governmental agencies about the effect of specific building features on fire suppression and occupant safety.
- the evaluation of potential heat and smoke damage for a particular building design.
- fire service participation in the global movement toward performance-based building codes with a rational, analytical method to assure that its "value added" to structural fire safety is properly recognized and an integral part of the process.

Size of a Fire that Can Be Extinguished

After the fire size at first water application can be estimated with some degree of confidence, the next consideration is the relationship between fire size, available suppression resources, and the selection of offensive or defensive attack modes. These tactics will influence the water requirements for control and extinguishment.

"Engine and Ladder companies work together: the ladder men begin ventilation as soon as the hose line is charged (i.e. water is up to the nozzle and ready to be applied). If ventilation occurs too

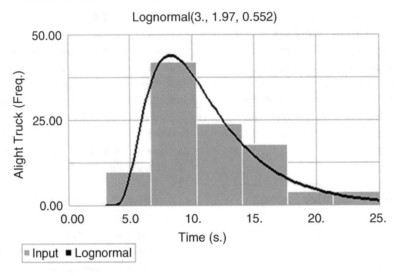

Fig. 8.2 Typical Firefighter Task – note the lognormal (not bell curve) shape of the distribution

soon, the air that is introduced will accelerate the fire (Warren et al. 1979) Firefighting in high-rise buildings is especially difficult for four reasons:

1. Aerial ladders will not reach beyond about nine stories: as a result, above these levels, the fire must be fought from the inside.
2. Complete evacuation of occupants is not feasible (therefore, provision for safety and life support systems must be made within the building).
3. Chimney or stack effects in tall buildings require provision for the control of smoke and products of combustion
4. Fixed windows make ventilating the fire floor difficult:

Average First Alarm Crews

Menke (Menke 1994) provides the following information on the number of hoses that can be put into operation and the size of fire in square feet that can be extinguished with each hose line. For Table 8.1: the question was "how many hose lines can effectively be put into operation for a high rise building".

For Table 8.2: the question was "for a ceiling height of 8′ to 10′, what is the largest fire that can be controlled and extinguished, by square footage?"

Table 8.1 Number of hoses put into operation

Hose size	1.5"	1.75"	2.5"
2nd Flr.	2	2	2
4th Flr.	2	2	2
8th Flr	2	1	1

Table 8.2 Largest fire controlled by one hose stream

1.5″ line	850 sq. ft
1.75″ line	1420 sq. ft
2.5″ line	1350 sq. ft

This information is based on a survey of 27 fire departments. The 2.5″ may be lower due to mobility issues.

References

Isman, Kenneth E. *Standpipe Systems for Fire Protection.* Springer, 2016.

Callery, James. "Building Evaluation for Manual Suppression." Worcester Polytechnic Institute, 1996.

Menke, Ken. *Predicting Effectiveness of Manual Suppression.* Worcester Polytechnic Institute, 1994.

Till, Robert. "A Building Evaluation Technique for Fire Department Suppression." PhD Dissertation. Worcester Polytechnic Institute, 2001.

Warren, Walker et al. *Fire Department Deployment Analysis – a Public Policy Analysis Case Study.* New York: Elsevier North Holland, Inc., 1979.

Sprinkler Systems and Their Types

9

Water Agent Suppression Systems

There is nothing new in the world, it has often been said: The same basic elements and concepts are merely used in different ways. This holds true for the wet pipe sprinkler system in the use of water as the agent in fire protection. All other automatic fire protection systems using water as the suppression agent have evolved from the wet pipe system. The wet pipe system has been adapted to specific requirements and environments, therefore, a thorough understanding of this system is the foundation on which to build basic knowledge of other systems using water as the fire suppression agent. After explaining the wet pipe system, this chapter will describe other systems in common use today: dry pipe sprinkler systems, pre-action systems, deluge systems, and water spray systems.

Wet Pipe Sprinkler System

The definition of a wet pipe sprinkler system in the National Fire Protection Standard 13 (2016) reads: "A sprinkler system employing automatic sprinklers attached to a piping system containing water and connected to a water supply so that water discharges immediately from sprinklers opened by heat from a fire". The wet pipe sprinkler system above ground is the consideration in this section, since the water supply and underground part of the system are covered in other sections.

The purpose of a sprinkler system is twofold: detection and extinguishment. Detection is accomplished when a sprinkler head fuses and water flows in the system, which activates a water flow alarm device which, in turn, activates the alarm system. The discharge from the fused sprinkler head provides the fire suppression.

Extinguishment is obviously with water. Wet pipe sprinkler system piping is constantly and completely filled with water under pressure, and when the sprinkler head is heated by a fire condition to the pre-set temperature, the fusible element releases the cap which encloses the water orifice, and water is immediately discharged onto the fire. This is one very important advantage of a wet pipe system—there is no delay in automatically applying extinguishing water after the sprinkler head has determined that the fire is of a magnitude requiring immediate suppression.

The piping in a wet pipe sprinkler system is sized to supply water, at the required volume and pressure, to a pre-determined number of heads. The number of heads anticipated to operate in a fire condition is based on studies and calculations of the fire spread in various occupancies and types of buildings. Chapter 10, Hydraulic Calculations, shows how the pressure at each sprinkler head is affected as more and more sprinkler heads become operable and their combined flow demands more and more water from the supply source through the supply piping to the operating heads.

If the fire should gain unanticipated headway, and the fire area extends beyond the anticipated, or calculated, area of fire spread, more sprinklers will operate than anticipated and more water will flow than the piping was designed to handle. As a result, pressure will drop, decreasing the individual sprinkler head discharge. With less pressure, the discharge pattern will not be within the design parameters of the sprinkler heads. The extinguishing efficiency of the sprinklers that are operating in the fire area will drop below 100% of design criteria and continue to drop every time another head operates outside of the anticipated area of spread.

As shown in Fig. 9.1, when a sprinkler (1) opens, the discharging water lifts the alarm valve clapper (2) and flows through the alarm port (3). When the retard chamber is filled, (4) water flows to the water motor alarm (5) and/or the optional pressure switch (6) which signals an alarm bell.

The wet pipe sprinkler system's strength lies in the fact that water is poised and ready to strike at the small fire, ready to shower its heat-absorbing, smothering, droplets of water directly on the heart of the potentially destructive combustion before the fire can intensify and spread.

Time is the greatest ally of a fire, which it needs as badly as it needs fuel and oxygen. The wet pipe system eliminates this time element by its speed in applying extinguishing water when the heat from the fire has reached the demand-for-attention point and actuated the sprinkler head directly over the fire.

Fig. 9.1 Wet pipe sprinkler valve. (Courtesy of the Viking Corporation)

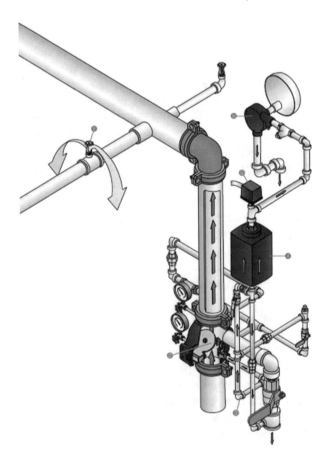

Advantages and Disadvantages

Before examining the components that comprise a complete wet pipe system, the advantages and disadvantages of this system should be evaluated.

The major advantage of a wet pipe system is that piping is constantly filled with water under pressure for fast discharge on a fire. The disadvantage is that the system can only be used to protect areas where the temperature does not fall below the freezing point. Only when a sprinkler head is open and discharging is there any flow of water in the piping, which makes this system very susceptible to freezing.

The fact that the water in the piping has no movement, except when the system is in an operational condition, helps prevent corrosion from building up in the piping. Corrosion occurs when a constant or occasional water flow passes through piping. (Corrosion in sprinkler piping is a saboteur working for the fire. It silently reduces the water passage area within the pipe by building up on the walls of the pipe, thereby cutting down on the amount of water that can pass through the pipe to the sprinkler head.)

In a wet pipe system, corrosion on the walls of the pipe is usually minimal since the volume of available corrosion chemicals in the water remains constant with no fresh supply of chemical available from flowing water. The temperature of the pipe and water assume identical levels in a relatively short time after initial installation and remain simultaneously constant. A change in room temperature between pipe and water will accelerate the deposit of scale or corrosion on the walls of pipe, but the stagnant water in a wet pipe system maintains a relatively constant pipe and water temperature, varying only if the room temperature changes.

As corrosion builds up on the walls of the pipe, the smooth surface gives way to roughness which, in turn, creates small eddies and turbulence as the water flows through the pipe. This turbulence slows water down, decreasing velocity of the water measured in feet-per-second (FPS) and allows additional pipe-surface/water-surface contact time. A slowdown of the water means more time for a pipe and water temperature differential to take effect, more time for chemical reactions to materialize, and more time for the corrosion process to have a maximum impact.

Wet Pipe System Alarm

When a sprinkler head is actuated and water discharges through the head, the sprinkler system has met two of the three required functions: detection and fire extinguishment or control. The third vital function is alarm.

A fire has been detected, and within the design parameters of the system, fire control has been initiated. As in any emergency, personnel are never sure whether or not their capabilities will meet the circumstances, and so they cry for help. In the case of the sprinkler system it may only be the need for the use of a hand extinguisher to snuff out the persistent fire under an obstruction that prevents sprinkler water from reaching the combustion zone. Or due to conditions that were not anticipated when the system was designed and installed, the sprinkler discharge can only hold an intense fire in check. In many cases, the professional assistance of firefighters with hose streams may be the only answer to complete extinguishment.

Water Damage

Consideration must also be given to water damage. With a wet pipe system, the water discharge commences immediately upon head actuation, which may eliminate an incipient fire within a few minutes, but the water continues to flow. At the rate of 15–30 gallons per minute, a few minutes can mean a lot of water discharged and subsequently a great deal of damage to an occupancy.

Sprinkler system alarm and life safety are inseparable in every type of occupancy, but particularly where life safety takes precedence over property protection. The sound of a sprinkler alarm not only demands attention in the fire area, but also initiates emergency procedures throughout a facility. Where the lives of incapacitated or restrained individuals are at stake, all the responsibility cannot be delegated to the ability of the sprinkler system; the alarm it has sounded is merely an indication that there is serious trouble and that the system will do its best to provide the time necessary to perform orderly emergency procedures.

Initiating the Alarm

There are two basic methods for the initiation of an alarm when a fire has automatically changed a dormant wet pipe system into a dynamic fire fighter, and both of these methods use the most obvious feature, flow of water, in this status change. The two methods are water flow indicator and the alarm valve.

Water Flow Indicator

A water flow indicator of the paddle or vane type utilizes a flexible vane of thin metal or plastic. A hole is drilled in either the horizontal or vertical sprinkler pipe that supplies all of the sprinkler piping and sprinkler heads on the system. Then the flexible vane is rolled up and inserted into the hole. After the rolled vane is entirely through the hole it unrolls or expands to its original shape, completely filling the pipe with the exception of a small working clearance between the vane perimeter and the inside pipe circumference.

The stem of this vane is connected to a mechanical linkage which is encased in a cover. When there is a flow of water in the pipe the vane is deflected, and the vane-stem actuates mechanical linkage which, in turn, makes an electrical contact. The vane's flexibility prevents the obstruction of water flow in the pipe, and when the flow ceases the vane returns to its original horizontal position.

The size of the hole that is drilled in the sprinkler pipe to admit the vane is dependent upon the pipe size. For pipe sizes 3″ and under, the hole

diameter is 1–1/4″ to 1–3/8″ (depending on manufacturer), and for larger pipe sizes the hole is 1–7/8″ to 2–1/8″ in diameter (again depending on the manufacturer).

Clearance around the pipe is required to allow free movement of the vane or paddle.

When the unit consisting of vane and alarm switches is properly located with the vane in the pipe and expanded, the unit is bolted to the pipe with a U bolt and tightened to ensure a leakproof seal over the hole.

Any flow equal to one sprinkler head discharging will actuate the vane-type water flow detector. On some wet pipe systems this one-head-flow-detection is expanded to include not only the flow from any head on the system, but also to pinpoint which sprinkler line or which section of the sprinkler system is in an operating condition. In addition to the water flow indicator installed on the main supply pipe, water flow indicators may be installed on a cross main supplying a section of the building, or even on individual sprinkler branch lines supplying but a few sprinkler heads.

By installing these sectional or zoned water flow indicators, the exact area, or zone, where a sprinkler head has operated will be indicated on an alarm panel, thus eliminating the search for which area of the system a head has actuated and is flowing. In a large office building or institution containing a multitude of small rooms, or in a multi-story building, the time lost in locating the operating sprinkler head can be very costly to life safety, water damage, or fire control. Practical examples sometimes help in evaluation of such an explanation. For example:

Scenario 1 Imagine that a six-story hospital with 50 patient rooms on each of four stories and the two remaining floors occupied by hospital support facilities such as operating rooms, storage rooms, and mechanical rooms, is completely sprinkled by a wet pipe sprinkler system with a vane-type flow indicator installed on the pipe which supplies the entire system. An alarm bell and light installed in the hospital telephone switchboard room and wired to the electrical contacts on the water flow indicator will be actuated by a flow of water from any sprinkler head anywhere on the entire system.

It is 2 a.m. and the telephone operator is startled by the fire alarm bell. She is a trained and experienced individual who immediately contacts the fire department to report a fire at Memorial Hospital! But where does she direct the fire department when it arrives? There have not been any in-house calls to report smoke or water, and the security and maintenance crew that were the second parties to be alerted, have begun a floor-by-floor search but have not reported to the switchboard any positive results of their search. Time at this point is on the side of the fire condition.

Scenario 2 Now consider the same hospital and the same 100% sprinkler protection with one significant difference. Each supply main that branches off the supply risers at every floor to supply the sprinklers on that individual floor are equipped with a water flow indicator. Now when the bell alerts the switchboard operator and a light on the annunciator is lighted behind the Floor-5 window, she will call the fire department and report-a fire, 5th floor, Memorial Hospital. The security and maintenance crews can be immediately sent to the 5th floor. Time is now on the side of the sprinkler system and the firefighters.

What has been described in Scenario 2 is a local sprinkler alarm. In Scenario 1, the bell alerts personnel on the premises, and someone calls the fire department. The human element is the weak link in this chain, and in a life safety situation, the need for professional help – the firefighters – is a vital element. With a local sprinkler alarm, the human link can be eliminated and the fire department summoned automatically when a sprinkler head operates, the flow of water begins, and the water flow indicator actuates.

The local alarm is not eliminated because the personnel at the facility must be alerted to a fire condition, so they can take all available first aid firefighting measures and initiate an evacuation

program, but an additional device is wired to the water flow detector alarm panel. An alarm transmitter which, when actuated, automatically transmits a signal over leased public telephone lines to a receiver in the fire station or fire alarm headquarters. This alarm receiver is marked for the facility it serves-Memorial Hospital Fire Alarm in this example. When this receiver is actuated, the fire department knows exactly where a sprinkler system is operating.

Every sprinkler system water supply is subject to water surges. This is a term which is self-explanatory. A wet pipe system is a tightly closed piping system. When the system installation is complete, and the water is forced into the system, some air is trapped in the ends of the lines where it is compressed by the water until the pressure of this small amount of air equals the water pressure. When there is a sudden pressure surge on the supply main, such as an additional pump cutting in on the city water system, a large flow of water downstream of the sprinkler system on the city water main supplying the sprinkler system is suddenly stopped by the rapid closing of a valve. This excess pressure seeks the area of least resistance, and since water is not compressible, the excess water pressure further compresses the trapped air causing a momentary flow of water in the sprinkler piping. This momentary flow will deflect the water flow indicator vane, operate the mechanical linkage, and the electrical alarm switch makes contact, and activates the alarm system. Since these water surges can occur with some frequency, especially on a city water system, this could mean many fire alerts and with a remote alarm system the frequent rolling of the municipal fire equipment. False alarms can be more dangerous than no alarm at all, since the old cry wolf can destroy the time advantage by waiting to check and see if it is a false alarm before alerting personnel and the fire department.

To control this unavoidable danger, water flow indicators are equipped with a retard feature. This retard feature literally retards the transmission of the electrical signal to the alarm system, and although it is factory-set, it can be adjusted in the field after installation. The retard element can usually be adjusted for a 0–70 s delay, which means the alarm signal can be transmitted the instant the vane is deflected, or the alarm signal transmission can be delayed for up to 70 s after the vane is deflected by a water flow.

This retard must have an instant recycling feature so that a sequence of flows, each of less duration than the predetermined retard period, will not have a cumulative effect. For example:

> A 30-s retard is actuated by a water pressure surge. Fifteen seconds into the retard, another surge occurs and now there is an accumulative retard of 45 s. Fifteen seconds later, another pressure surge occurs and now there is an accumulative retard of 75 s. With the instant recycling feature, the first retard time remaining would have been canceled when the second surge occurred causing a new 30 s retard, and so on with the subsequent surges. Adjusting the retard must be accomplished by experienced personnel with the approval of the authorities having jurisdiction, but the instant recycling feature must always be specified when a flow switch is part of the specification.

Alarm Valve

The water flow indicator functions electrically. Therefore, if there is a power failure and the facility does not have a back-up power system or the alarm system is not equipped with a battery back-up, a flow of water in the sprinkler system will not produce an alarm. Primarily through the influence of insuring interests, with their concern for property protection and their fear of excessive water damage from undetected sprinkler discharge, most wet pipe systems are equipped with an alarm valve.

The alarm valve, often referred to as a glorified check valve, is a special check valve with water flow detecting capabilities. This device not only detects flow, but can operate a mechanical local alarm, a water motor alarm, or activate an electrical alarm device.

The alarm valve is installed in the main sprinkler supply line close to the point where the underground water supply enters the building.

The alarm valve is installed in the pipe that supplies water to the entire sprinkler system so that a flow of water from any operating sprinkler

head on the system must pass through the alarm valve.

All alarm valves are different in some respect, but the basic design and function of all alarm valves is identical. The alarm valve body is cast iron with standard flanged or grooved connections. It contains a clapper similar to a check valve, with a rubber or neoprene seat where it makes contact with the bronze water passage opening.

Being a wet pipe system, the piping and alarm valve are completely filled with water, and the pressure will be equal on the top and bottom, supply and system side of the clapper position. If water surges have occurred which force excess pressure through the alarm valve compressing the air in the sprinkler line piping, the clapper will open momentarily to admit the excess pressure, and the close again when the surge recedes, trapping the higher pressure on the system side of the clapper. For this reason, there are two different pressure gauge readings on an alarm valve. One gauge monitors water pressure on the supply side of the alarm valve clapper, and the other monitors rate pressure on the system side of the alarm valve clapper. This difference is not objectionable since it holds the clapper in the closed position and prevents the possibility of false alarms. This will become clear in further examination of the alarm valve function. When a sprinkler head opens and allows water to discharge, there is a steady flow of water through the alarm valve which raises the clapper. While the clapper is in the closed position, it seals off an opening to the alarm line. When the clapper raises, the water pressure forces water into this small alarm line pipe opening. This feature separates the alarm valve from the standard check valve, and justifies calling this piece of equipment an alarm check valve.

A slight amount of excess pressure on the system side of the alarm valve clapper diminishes the chance that the clapper will raise with a surge of supply pressure and expose the alarm line opening. This momentary raising of the clapper will occur from time to time, and water will be admitted to the alarm line, so provisions must be made to prevent false alarms. Like the water flow indicator, alarm actuation must be retarded to prevent false alarms.

Water Motor Alarm

The water that enters the alarm line, although a small stream, has the same pressure as the sprinkler system water supply, and this pressure is used to operate a mechanical water motor alarm. This alarm device consists of a paddle wheel in a sealed housing. The pressurized water stream from the alarm line discharges against the fins or paddles of the water wheel and mechanically turns the center shaft on which the paddle wheel is mounted. On the other end of this shaft is mounted a striker which when rotated within a metal bell casing continuously strikes or rings the bell.

The water motor, or paddle wheel, is usually installed on the inside wall of a building with the shaft extending through the wall in a shaft sleeve, and the bell mounted on the exterior side of the wall. This is strictly a local alarm for the wet pipe system, and the efficiency of the water motor as an alarm is only as great as its sound range, and personnel within that sound range. It is a back-up, a simple, reliable, mechanical, water flow indicating alarm. If it is required, or desired, to supplement this local alarm with an electrically operated water flow alarm, a device known as a pressure switch is also used in conjunction with, or in lieu of the water motor alarm, if approved by the authorities having jurisdiction.

Pressure Switch

The pressure switch on an alarm valve is also activated by water pressure transmitted through the alarm line. This pressure acts on a diaphragm in the pressure switch, forcing it to move and thus make electrical contact. The details of the pressure switch will be further described in the section on alarm devices.

Retard Chamber

System water pressure transmitted through the alarm line is the force which operates both the mechanical and electrical alarm devices on a wet pipe sprinkler system. The next device which warrants consideration is the retard chamber which prevents water surges from producing false alarms.

The alarm line runs from the alarm valve to the retard chamber and from the outlet on the retard chamber to the water motor alarm and/or the electric pressure switch. The retard chamber is merely a holding tank, having a capacity of approximately two gallons. The inlet to the retard chamber is through a restricted orifice approximately 5/16″ in diameter attached to the 3/4″ alarm line from the alarm valve. This orifice restricts or slows down the volume of water which the water pressure in the alarm line can force through the opening and into the retard chamber. In other words, it restricts the flow of water to the retard chamber.

The retard chamber must completely fill with water before it can discharge through a 3/4″ opening at the top of the chamber. Water continues on through the alarm line to actuate the water motor and/or force the diaphragm in the presence switch to deflect. This makes an electrical contact to complete a circuit, light lights, ring bells, or actuate a transmitter to automatically conduct a signal to a remote alarm. When the wet pipe sprinkler system is subject to abnormally large water surges, it may be necessary to install two retard chambers in tandem. The 3/4″ outlet at the top of the primary retard chamber is piped into the drain opening in the bottom of the secondary retard chamber, and the outlet to the alarm devices is connected to the alarm line at the top of the secondary chamber. Both retard chambers must completely fill with water before the water pressure can be transmitted to the alarm devices.

At the base or bottom of the retard chamber is a small 1/8″ restricted orifice drilled into a 3/4″ drain line. If a water surge has passed water into the alarm line and thence to the retard chamber, and has half filled the chamber by the time the alarm valve clapper resets and shuts off the alarm water, the water in the retard chamber will drain out slowly through the 1/8″ restricted orifice into the 3/4″ drain pipe (which is usually connected to the main drain pipe from the alarm valve).

There are retard chambers with more sophisticated entry and drain mechanisms. The retard chamber with a clapper and mechanical linkage arrangement is one example. When water is introduced into the retard chamber from the alarm line, it forces a small clapper to open the inlet waterway, and the clapper (being connected by an arm to a diaphragm) automatically closes the drain opening. When the water pressure on the alarm line is relieved by the alarm valve clapper resetting and closing the alarm line outlet, the retard chamber inlet clapper resets due to a spring mechanism and subsequently releases pressure on the drain diaphragm. This, in turn, opens the drain opening and the false alarm water in the retard chamber drains off into the drain line.

If the alarm system consists of a retard chamber and pressure switch only, it will be necessary to bleed-off the air in the retard chamber since the pressure switch is connected to the 3/4″ outlet from the top of the retard chamber. The entire complex, with water covering the chamber drain opening, is air-tight. As the water rises in the chamber, the air will be compressed until it reaches the pressure of the water. At this point of equalization, no more water will enter the chamber. The pressure switch has a very tiny air-compensating vent which, under normal conditions, maintains an equal air pressure on both sides of the electric contact diaphragm. This compensating vent may not bleed off enough air pressure to prevent this air/water pressure equalization, and this can very possibly prevent the diaphragm from receiving sufficient alarm water pressure to deflect sufficiently to make the electric alarm contact. The water pressure acting on the diaphragm is necessary to cause sufficient movement for a positive electrical contact. To alleviate this condition, a line containing a restricted orifice is installed between the retard chamber and pressure switch. This line discharges into the main alarm valve drain line and bleeds off the excess air

pressure. If this bleed-off arrangement is not provided, and a water motor alarm (which bleeds-off pressure) is not installed, an automatic air vent on the retard chamber must be installed.

Alarm Valve Trim

The standard trim for an alarm valve consists of two pressure gauges to monitor supply and system water pressure, and each gauge is equipped with a small (usually 1/4″) control valve so that a deflective gauge may be replaced without disturbing the entire system. Each pressure gauge line has a 1/4″ plug which can be removed, providing an opening for an inspector to install his pressure gauge to check the accuracy of the installed gauges.

Taking suction from the supply side of the alarm valve clapper is a small alarm test line piped through a bypass test valve, which is normally closed and only used for testing the alarm system. It is then piped to the retard chamber. The inspector can open this bypass test valve and actuate the alarm system with a simulated flow from the alarm line. The retard chamber can also be removed for repair or replacement without shutting off the system water supply by installing an indicating control valve, which visually indicates the open and closed position between the alarm line opening under the alarm valve clapper and the retard chamber.

The wet pipe system main drain is connected to an opening in the alarm valve body above the clapper and is piped through an angle control valve to discharge through the building wall to the atmosphere or to an open floor drain, roof drain, or other discharge receptacle that can accept the discharge flow. The main drain can be of more importance than merely draining off the system water when it becomes necessary to make repairs to the system. The same holds true when the area protected by the system will be subject to freezing temperatures and the water must be drained out of the system. If a sprinkler head should be damaged (hit by a lift truck, a rolling scaffold, a ladder, etc.) and commence discharging water, the first action taken to avoid serious water damage is to close the main control valve to shut off all water supply to the system. If the main drain is simultaneously opened wide, a great deal of system water can be drained off through the main drain instead of running off through the open sprinkler head. This procedure will immediately relieve some of the back pressure which is forcing water out of the open-head even after the main water supply is shut off. This is one reason why the main drain discharge must be unrestricted and must discharge where the full flow of the drain line will not cause additional water damage. The size of the drain line is dependent upon the size of the system. The size of the system riser supplying the system determines the system size:

A 4″ or larger system requires a 2″ drain, a 2–1/2–3″ riser needs a 1–1/4″ drain, and smaller risers require a 3/4″ drain.

Fire Department Pumper Connection

The fire department pumper connection, or siamese connection, is a common sight on the exterior wall of a building. The Y-shaped pumper connection has caps covering the two openings of the 2–1/2″ pipes which protrude from the casting that is the body of the Y. On buildings where appearance is a factor, all that need be seen are two, usually chrome-plated, caps installed on concealed 2–1/2″ pipes projecting through an oblong metal wall plate. These wall type fire department pumper connections are commonly sized 2–1/2″ by 2–1/2″ by 4″; meaning that the two 2–1/2″ outlets are siamesed into a 4″ casting into which is screwed a 4″ supply pipe.

This 4″ pipe is piped into the main sprinkler riser-the pipe that supplies all the water to every head on the system. (The diameter is used for example only. Under certain conditions, a 6″ or 3″ pipe may be used.) As soon as the fire department arrives at the scene of a fire, they will connect a line from their pumper truck to the closest hydrant. From two pumper truck discharge outlets, two 2–1/2″ lines will be connected to the two 2–1/2″ outlets on the sprinkler system sia-

mese connection. The fire department can now pump water at a high pressure and volume directly into the sprinkler system. If the fire has overpowered the sprinkler system (more heads have opened than were anticipated in the hydraulic design of the system) this water being pumped into the system can augment the volume and increase the pressure to overcome the increased demand of the discharging sprinklers.

The records of the National Fire Protection Association (NFPA) are a tragic reminder of how often a main water supply control valve to a sprinkler system is not reopened after a system repair is completed. In these cases, the fire department pumper connection is the only source of water supply to the firefighting sprinkler heads. For this reason, the fire department pumper connection must always be connected to a sprinkler riser on the system side of all control valves.

The pipe from the siamese connection to the sprinkler system must always have a check valve through which the fire department can pump water into the system, but when not in use, the system water pressure will not leak out, or flow out through the siamese connection when caps are removed, and hoses connected. In cold climates where the siamese connection on the exterior wall of the building is subject to freezing, the check valve keeps the line to the siamese dry and empty of water.

To drain off water in this section of line after the siamese connection has been used, to drain off any leakage of water through the check valve, and to drain off condensation, an automatic ball drip drain is installed on the siamese side of the check valve either at the check valve or at a low point on this segment of piping. One type of automatic ball drip consists of a small, usually, cast iron body drain containing a solid metal ball which rests in a depression beside the open drain opening. When the ball drip drain is under pressure while the siamese is in use, the water pressure forces the ball over the drain opening thus closing off the drain.

One additional element must be considered. Suppose the sprinkler system water supply has a pressure of 60 pounds per square inch (psi) and is fed from a city water main. If the fire department pumps into the siamese connection at 100 psi with a capacity of 500 gpm, and the sprinkler heads that have opened are only discharging a total of 200 gpm, the water will seek the path of least resistance, the 60 psi water supply. This will force the supply water back into the city water mains, completely nullifying the Fire Department siamese water and pressure to the open heads on the system.

Every sprinkler system must have the auxiliary water supply afforded by the siamese connection. In addition, every fire department siamese connection inlet pipe must have a check valve to prevent system water discharge back valve through the siamese. Every sprinkler system must have a means of checking the water supply piping to prevent flow back to the supply source when the fire department is pumping into the siamese. When a vane or paddle type water flow indicator is the alarm device on a wet pipe system, a check valve must be installed in the main sprinkler water supply on the water supply side of the connection of the siamese. This is connected to the sprinkler riser to check the flow of the siamese back into the supply.

When an alarm valve is the water flow alarm device on a wet pipe system, the check valve feature of the alarm valve can be used to accomplish this supply check. The siamese supply is connected to the sprinkler system on the system side of the alarm check valve. The siamese connection on the exterior wall of the building should be located not less than 18″ nor more than 5′ above grade (or the access level of the Fire Department).

Main Control Valve

The main control valve is installed in the main water supply pipe to the sprinkler system, and controls all of the water to the sprinkler system. The control valve may be installed inside the building where the underground supply pipe enters the building, or it may be installed outside the building in the underground supply pipe. Inside or outside, all sprinkler systems, and in fact all fire protection control valves, are required to have one feature-visual indication of an open or shut position. This section will consider only the inside, or above ground, control valve.

The terms inside or outside control valves refer to valves installed in the underground pipe outside the building, as opposed to valves installed in the steel pipe of the sprinkler system above ground, and inside the building. A commonly used inside control valve for sprinkler systems is the Outside Screw & Yoke (OS&Y) valve.

This meets the requirement for all fire protection valves controlling water supplies; namely, a visual inspection will tell at a glance whether the valve is open or closed. As a result, all valves controlling water supplies must be indicating type valves, which means they must visually indicate whether they are open or shut.

There are several indicating type control valves on the market for use on fire protection systems. Except when the use of an OS&Y valve is required by NFPA or the insurance carrier, butterfly control valves may be considered to save space.

The control valve on a sprinkler system serves several basic functions. This is necessary in order to shut off the water supply to the system so that the system can be repaired, altered, or increased in size.

When the sprinkler system has suffered physical damage, and is discharging water through a broken pipe, fitting, or sprinkler head, the supply of water to the entire system can be stopped by closing the main control valve.

When the sprinkler system has performed within the efficiency of the design parameters and extinguished a fire, the water damage can be kept to a minimum by closing the main control valve and stopping the water supply to the heads that have been opened by the fire. At this point, it would be inexcusable not to mention the tragic record of major fire losses that have resulted from this misuse of the main control valve. A fire that produces enough heat to operate a sprinkler head cannot be safely declared officially extinguished by the average layman. All too often in their haste to avoid water damage, a main control valve will be prematurely closed because the fire appeared to be out. Control valves are not always conveniently located within eyesight of the fire. By the time the individual who has been dispatched to the basement where the valve is located returns to the third floor where the fire has occurred, he finds that the fire has re-ignited and is using the time for his return trip to reopen the valve to feed itself, uninterrupted by the sprinkler discharge. The time element involved has changed the complexion of the fire situation from a controlled to an uncontrollable fire condition.

Many NFPA fire records have shown that a control valve was shut off too quickly, and when the error was discovered the fire had made it impossible for an individual to approach the valve location to reopen the control valve.

Without a water supply, a sprinkler system is nothing more than a complex of pipe and fittings, which points out another facet of the importance of the open control valve.

> When sprinklers operate, they are effective 97% of the time, resulting in a combined performance of operating effectively in 91% of all reported fires where sprinklers were present in the fire area and fire was large enough to activate them. The combined performance for the more widely used wet pipe sprinklers is 92%, while for dry pipe sprinklers, the combined performance is only 79% (Hall 2010).

The fire department uses the sprinkler system control valve in situations where, in their professional opinion, the fire has overpowered the sprinkler system. So many heads have opened, and so much water is being discharged, that it is affecting the water supply and pressure at the fire hydrants. The sprinkler system has lost its value at this point and has in turn become a deterrent by robbing the hydrant hose streams of vital volume and pressure, and the fire department will close the control valve. Another case where the control valve is shut occurs when there is a flash fire: When an ignition source ignites fumes, which have spread over the entire floor area, a massive flammable liquid spill covers a wide area, or the preheated combustible construction of an entire room. The flash of heat can open heads far

removed from the primary fire source, and is usually intense enough to fuse sprinkler heads, but not of sufficient duration to sustain continuing combustion within its area of influence. The result will be a hundred sprinklers operating on a system calculated for the operation of 30 sprinklers, and the drain on the water supply, both pressure and volume, far exceeds design parameters. The fire department will also close the main control valve to save water for their hose streams when a sprinkler system supply pipe has been broken by an explosion, collapsing wall, or structural member. The location of a main control valve on a wet pipe system incorporates what we have learned about the uses of a main control valve and the fire department pumper connection. The control valve on a wet pipe sprinkler system is always installed on the supply side of the fire department pumper connection. By following this rule, a closed control valve can never prevent or obstruct the use of the fire department pumper connection.

Inspector's Test

The wet pipe sprinkler system has been installed, and by all appearances is ready to perform its designed function. Unlike almost every other building system, however, its operational ability will not be demonstrated on a daily basis. The sprinkler system may not be called upon to function for 10 or 20 years, but when it is needed, the system must function without fail. The eruption of fire is no time to discover a mechanical defect that affects the system operation, and sprinkler systems are seldom afforded a second chance.

In order to test this completely closed and sealed system without the expensive and impractical fusing of a sprinkler head, each wet pipe system is equipped with an inspector's test connection.

The inspectors test connection consists of a 1″ valve installed in a 1″ pipe that takes supply from the wet pipe sprinkler system. When this valve is opened, water is discharged through a smooth bore corrosion-resistant outlet or bushing that produces the flow equivalent to the flow from one open sprinkler head. Opening the inspectors test valve will, if the system is designed, installed, and maintained properly, operate all of the wet pipe system equipment and actuate all alarm equipment. In order to obtain a true reading on the system, the inspectors test should be installed at the end of the branch line that is most remote from the system supply, and at the highest point on the system. If the supply pressure is not sufficient to provide adequate volume and pressure, the inspectors test will not have sufficient pressure to discharge an adequate amount of water to operate the system or alarm equipment. The inspectors test is a valuable tool for the experienced inspector.

Discharge

The discharge from the inspector's test must be unobstructed and visible. Discharge may be either to atmosphere or, where this is not practical, the test pipe may terminate in a roof or non-sanitary drain pipe capable of accepting full flow under system pressure. A sight test connection containing a smooth bore corrosion-resistant orifice providing a flow equivalent to the flow from one sprinkler is installed between the inspector's test valve and the connection to the drain line.

Location

The test valve should be located where it will be readily accessible and not over 7′ above the floor. In the discussion of water flow indicators, the use of these devices at various locations on a wet pipe sprinkler system to afford pinpoint location of a water flow requires the installation of an inspector's test connection to facilitate testing each of these water flow detection devices.

Where the test connection to operate a vane-type flow switch cannot discharge to the exterior of a building and there are no suitable plumbing drains to accept the test discharge, a closed system test and drain may be used. For example, a multi-story dormitory completely protected with

a wet pipe sprinkler system: In order to have an alarm panel that indicates which floor has a sprinkler head flowing, the panel is wired to a flow switch on each floor, installed in the pipe supplying the sprinkler on the floor. The supply to the sprinkler on each floor of the dormitory is controlled by an indicating control valve. Should a leak have to be fixed, or a sprinkler head replaced, the floor control valve can be closed, isolating that one floor sprinkler system and leaving all other floors under sprinkler protection. In order to be sure that the water flow switch on each floor is working properly, and to drain the piping once the control valve is closed, a test and drain set-up can be installed on each floor.

The test and drain set-ups discharge into a dedicated drain riser, which picks up the test and/or drain water at each floor and carries it to an acceptable location at ground floor level for discharge.

Wet Pipe System Drainage

References have been made to shut the main control valve and drain the system for maintenance, alterations, or to freezing temperatures. The ideal design of a wet pipe sprinkler system would be to have all of the water in the piping network drain back to the main drain, but due to changes in elevation, there are usually low points, trapped sections of piping, or heads on the system requiring auxiliary drains. To quote from the NFPA 13 Standard for the Installation of Sprinkler Systems, all sprinkler pipe and fittings shall be so installed that the system may be drained. (8.16.2.2.2 -2016).

NFPA is a requirement to accomplish wet pipe sprinkler system drainage was to pitch the pipe so that all water, exclusive of trapped areas, would automatically drain back to the main drain. This rule has since been changed so that now wet pipe sprinkler piping may be installed level. It was determined that the inherent pipe size change in sprinkler piping with a level center-line would cause the water to naturally flow from the small to the large diameter pipe, and thence back to the main drain. This change has been a tremendous sprinkler design advantage where long runs of pipe are coupled with close clearance tolerances, such as above a suspended ceiling. The requirements for draining trapped sections of a wet pipe sprinkler system are based on the number of heads that are trapped or the number of gallons of water that will drain back to the main drain by gravity. These auxiliary drains have no bearing on the fire protection efficiency of the system and are installed to minimize water damage in draining the system.

A drain plug in a small diameter pipe creates the impression that only a small amount of water is trapped in the piping. Sometimes due to construction or occupancy, sprinkler piping is not readily visible. It does not take much imagination to visualize what happens when a workman removes a 1″ plug and holds up a 5-gallon bucket to catch the water, not knowing that the 1″ pipe is on the end of a large trapped section of sprinkler piping containing 30 gallons of water.

Where it is obvious that the trapped section contains less than five gallons, a drain plug is all that is required. However, when a trapped section appears to contain several gallons, supplies quite a few sprinkler heads, or has a rather large pipe supply, the installation of a 1″ drain valve will facilitate draining. The ideal situation is to discharge a drain line to the exterior of the building.

The wet pipe sprinkler system is simple, efficient, and economical to install and maintain, accounts for the majority of all sprinkler installations. Dry pipe sprinkler systems, on the other hand, account for a much smaller percentage of sprinkler systems in use. They are designed for use in facilities in which it is either impractical or impossible to provide the heated environment necessary for the proper functioning of a wet pipe system. The dry pipe system provides fire protection to unheated structures, and in some cases, specific unheated areas of the structure in which the system piping is subjected to freezing temperatures. Use of a traditional wet pipe system in an unheated zone would result in frozen pipes, which could block the flow of water to a fire or cause considerable water damage should the pipes burst. This section examines the basic oper-

ation of the dry pipe sprinkler system, including advantages and disadvantages of the system and an explanation of the systems components.

Dry Pipe Sprinkler System

The dry pipe system, like the wet pipe system, uses water as the extinguishing agent. Unlike the wet pipe system, however, the dry system piping is filled with pressurized air or nitrogen. When a sprinkler head fuses, the air or nitrogen pressure is dissipated through the open head. The piping then fills with water and discharges through the open head. Understanding how a dry pipe system operates requires a brief lesson in physics. A moderate amount of pressure on one side of a large surface can equalize a greater amount of pressure on a small surface on the opposite side. The combined pressure on the total surface area can be found by multiplying the amount of pressure expressed in pounds per square inch (psi) by the number of square inches.

Most dry pipe valves are designed so that a moderate amount of air pressure on the larger diameter air side of the clapper in the dry pipe valve will hold back a much higher water pressure acting on the smaller surface water side of the clapper. (See Fig. 9.2). When the air pressure can no longer contain the water pressure, the valve will open by raising the clapper. The difference between the air pressure and the water pressure at this point is expressed as a ratio. This ratio is called the differential. Dry valves that operate on this principle are called differential-type dry pipe valves.

Basically, dry pipe sprinkler system piping is filled with pressurized air or nitrogen. When a sprinkler head operates, the compressed gas is discharged faster than the systems compressed air source can replenish it. The pressure on the air side of the clapper drops to the valve trip point, which is the point at which the water pressure overcomes the air pressure on the air side of the clapper. The clapper flies open and water rushes through the system to the open sprinkler head(s).

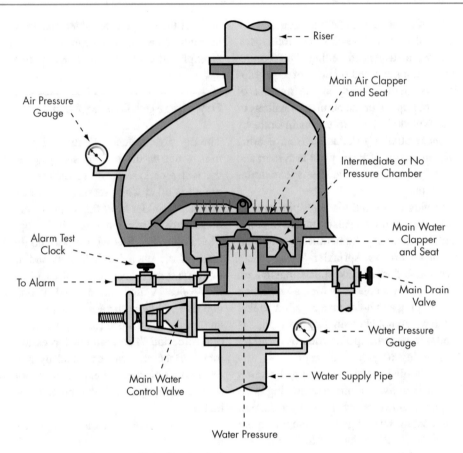

Fig. 9.2 Dry pipe valve. (Courtesy of John Morrison)

Advantages and Disadvantages of Dry Pipe Systems

The basic advantage of a dry pipe system is that it can be installed in areas subject to freezing. Dry pipe sprinkler systems are an economical way to provide sprinkler protection in unheated areas. Unfortunately, there are drawbacks to this system.

One disadvantage of this design is the time lag inherent in the operational sequence of a dry pipe system. This is time that enables the fire to intensify and spread. When the sprinkler head is actuated by the heat of a fire, and the sprinkler head discharge orifice is opened, it is necessary to reduce the air pressure that fills the system to the trip point by releasing this air through the sprinkler orifice. It is not unusual to have a dry pipe system containing 300 or 400 gallons of air at 30

or 40 pounds of pressure. With standard orifice sprinkler heads, this can mean a considerable time delay to evacuate a sufficient volume of air through the orifice to reduce the pressure to the trip point. Once the water is released into the system, it must overcome the back pressure of the remaining air in the piping in traveling to the open sprinkler head. While sprinkler head actuation-to-water discharge time in a wet pipe system is instantaneous, with a dry pipe system it may take 60 s or longer, depending on the size of the system and the air pressure level. This is precious time that could be used extinguishing the fire.

NFPA records indicate that on the average, more sprinklers are actuated in areas protected by a dry pipe system than in areas protected by a wet pipe system (because the fire is bigger due to the time lag). The designed area of application for a wet pipe system is increased 30% to compensate for this – hence the opening of more sprinklers is accounted for.

Another disadvantage of the dry pipe system is corrosion. The piping is filled with compressed air supplied to the system from one of two sources-air compressor or shop air. Both of these air supply methods obtain their air supply from the atmosphere, and at ambient temperature. The air supply may not be the same temperature as the piping of the dry system, and if relatively warm air is pumped into dry system piping installed in a cold area, condensation of the moisture in the warm air will collect on the interior walls of the system piping. This creates an ideal atmosphere for the formation of corrosion or scale in the piping.

Aside from producing corrosion, condensation causes another problem. Where the temperature differential is severe (for example, where a dry system is installed in a walk-in freezer and the dry valve and air compressor are located outside of the freezer) the smaller diameter piping can fill with water, freeze, and can crack the pipe. For this reason, the system air supply to the compressor must maintain a temperature closely approximating the temperature of the area where the piping is installed. Condensation and corrosion can be fairly well contained, except under severe conditions. For example, a dry pipe system protecting an unheated warehouse where the temperature of the piping under a metal roof may reach 100 °F in the daytime and drop to 60 F at night will be particularly vulnerable.

Differential Dry Pipe Valves

The differential-type dry pipe valve has a trip ratio of approximately 6:1. In other words, the valve operates when the air pressure in the system is reduced to approximately one-sixth of the system water supply pressure. For example, in a differential dry pipe valve with a 6:1 ratio, operating on a system water supply pressure of 100 psi, the trip point would be 100 divided by 6, or 17 psi air pressure. Standard practice is to maintain an air pressure 15–20 psi above the trip point to reduce the danger of accidental tripping

when the water pressure is relatively high, or the water supply is subject to surges. Several devices have been perfected to narrow this time lag.

One such device is the differential dry pipe valve. Naturally, dry pipe valves cannot be manufactured to accommodate different water pressures. A 6″ dry pipe valve may be installed where the water pressure is 60 psi, or the same valve may be installed where the water pressure is 125 psi. The differential of the particular valve is known, and the air pressure must be adjusted to meet the water supply pressure. To accommodate the relatively high water pressures that could be encountered, the air side of the clapper has to be large. Subsequently, the body housing these large clappers became rather massive in comparison to the size of the alarm valve.

Low Differential Dry Pipe Valve

The low differential dry pipe valve has a differential in the range of 1.0 and 1.2–1 instead of the 5: or 6:1 ratio common with the ordinary differential dry pipe valve. The low differential dry pipe valve resembles an alarm check valve both in size and operation. The air pressure on the sprinkler system side of the clapper (the air side) actually exceeds the pressure on the water side of the clapper. The valve operates when the air pressure is reduced to only 10% less than the water pressure. With a water pressure of 100 psi, for example, the air pressure would be 115 or 120 psi. To operate the low differential dry pipe valve, the air pressure would only have to drop to approximately 10% below the water pressure (in this instance 90 psi). With air pressure this high, bleeding off 20 or 30 psi through an open sprinkler orifice would require only a few seconds, unlike the excessive trip time of ordinary differential dry pipe valves.

The low differential dry pipe valve was found to have an advantage beyond its smaller size. An ordinary dry pipe valve provides an unobstructed flow of water when the valve is tripped. With the conventional 5: or 6:1 differential-type dry valve, using low system air pressures as compared to the supply water pressures, the rush of water into the system is tremendous, since the low air pressure offers little compressive resistance. In the low differential-type dry pipe valve, on the other hand, the system air pressure at trip point nearly approximates the supply water pressure, so that the velocity of water entering the system is greatly reduced. Aside from reducing the possible damage to the sprinkler system caused by a pressure surge of high-velocity water through a tripped dry valve, the amount of debris, corrosion, scale, and sediment entering the system is reduced with a low-differential dry pipe valve because the carrying force of the low-velocity water is greatly diminished. This is crucial to the effectiveness of the system, since scale and sediment carried to the open heads can clog the sprinkler head orifices and impede water discharge during a fire.

Note how similar the low differential dry pipe valve is to the alarm valve, not only in appearance, but in the method of charging the alarm line. A pilot valve on the clapper arm closes the alarm line opening when the clapper is in the closed position. When the valve is tripped, the clapper is raised and the alarm line orifice is opened. Note, too, that like the alarm valve, the clapper is not latched in the open position, but is free to swing shut.

It is not accepted practice to use this valve as a check valve opposing the fire department pumper water pressure because of the possibility of damaging the rubber clapper ring that seals the clapper. As a seal, this clapper is important regardless of the type of valve. The fire department pumper connection to the dry system must be made on the wet side of the dry pipe valve due to the inability of the check valve in the fire department pumper line to hold air pressure, and the inability of the dry valve to act as a check valve.

In the dry pipe valve, special rubber valve rings are used on clappers to maintain a tight seal when the clappers are in the closed position, and to compensate for any minute metal pocket, pitting, or minor metal-to-metal dimensional variations that could allow air to escape. As a further safety factor, the air side of the clapper is primed with enough water to completely cover, and thereby seal off, the clapper, further preventing

air loss. This priming water also keeps the rubber seals from drying out and shrinking or cracking.

Dry System Air Supply

Whether a dry system has its own air compressor or uses shop air, the air supply to the system must be regulated to maintain the required system air pressure.

When a dry system has its own air compressor, this operation is performed by a hi-low switch. The hi-low switch is an air pressure operated electric switch connected to the dry system air supply. When the system air pressure drops below a predetermined point (still well above trip pressure), the pressure switch is activated and closes an electric contact which automatically starts the air compressor. This, in turn, supplies air to the system until the system pressure reaches the required pounds per square inch. The preset pressure again activates the hi-low pressure switch, and the electrical contact is broken, stopping the compressor.

The air supply for a dry system should be capable of restoring required air pressure in the system within 30 min in ordinary differential (5: or 6:1) dry pipe valve systems, and within 60 min in a low differential dry pipe valve system. The compressed air supply should be provided with at least one relief valve set to operate at a pressure 5 psi in excess of the maximum system air pressure.

Shop air (compressed air used in operating the plant) is not always within the acceptable pressure range of that required in a dry system, and due to facility use, often fluctuates. However, initial equipment and maintenance costs can be eliminated by using shop air in lieu of providing and installing a separate air compressor. The installation of a relatively inexpensive and maintenance-free device is an existing air maintenance system that can provide acceptable control of shop air. The air maintenance device is installed in an airline from the shop air system to the dry system, where it mechanically and automatically accepts the pressure the shop air delivers, and maintains the constant required air supply pressure to the dry system.

A note of caution: The pressure maintenance device will accept shop air pressure that is higher than the air pressure required in the dry pipe system piping. The device will automatically convert higher pressure to the lower constant pressure required for the dry pipe system but cannot accept a shop air pressure lower than the dry system pressure and convert it to the required dry system pressure. In order to provide a continual acceptable pressure for fire protection, shop air must be maintained at a higher pressure than required for a dry pipe system. Another factor to monitor is during operation, ensure that the shop air used in the plant does not pull the pressure below the dry pipe system requirements.

Other maintenance considerations include plant shut down. When the plant is shut down for weekends, holidays, or at night, the shop air compressor cannot shut down; another condition where this could become a problem would be during a strike.

If shop air is not a reliable source of pressure, the next choices are an air compressor dedicated to supply air to the dry pipe system or a small 1/4 horsepower maintenance compressor to supplement the shop air.

The automatic air supply to the dry system cannot in any way interfere with the operation of the dry pipe valve by replenishing the air supply when a sprinkler head has operated and is releasing the air pressure. The air from the air supply is admitted to the system through an orifice which is so small that it cannot possibly admit enough air quickly enough to interfere with the tripping of the dry pipe valve.

Optional Devices

On large dry pipe systems, where temperature changes may produce periodic fluctuations in system air pressure, and where the possibility of the dry valve tripping could cause very serious problems, an automatic air compressor may be equipped with a reservoir tank. The air compressor maintains a constant air pressure in the tank and this tank air pressure floats on the dry system, which prevents the frequent cycling of the air compressor and

increases the reliability of constant air pressure in the system. The tripping of a dry pipe valve due to low air pressure and the filling of the piping with water in an area subject to freezing is always a serious problem if not detected before the water freezes and cracks a fitting.

This problem quickly goes from serious to critical when this occurs with a dry pipe system protecting a freezer. The water will not only freeze very quickly, but if ice forms in the piping, and the freezer cannot be shut down until the piping thaws, it may even become necessary to disassemble the system, thaw out the piping, and reinstall the system, obviously a costly operation.

Inexpensive devices can be installed on dry pipe systems to protect it from dangerous air pressure fluctuations by initiating an alarm, such as a low air trouble alarm. Automatic air supplies are dependent on mechanical and electrical equipment which is naturally subject to failure. This failure potential is usually in direct proportion to maintenance of the equipment. If the automatic air supply to the system should fail and the system air pressure drops below a predetermined point, it would actuate the low air alarm pressure switch, electric contacts would close, and a trouble alarm would alert personnel to the need for immediate attention before the pressure drops to the trip point. These devices should supervise the dry system air pressure and activate a trouble alarm for either a predetermined low or high pressure setting. The air supervisory switch should be included when specifying a low air alarm device.

When the device is activated, and especially when it is activated by low air pressure, the trouble signal must be automatically transmitted to a constantly attended location. A low air trouble alarm means the water supply control valve for the dry system must be closed immediately until the problem is corrected to prevent serious damage to the system and/or costly maintenance.

Another optional device that may be required is a dehydrator on the air supply to the dry pipe system. The dehydrator is used to dry the pressurized air supply to the system to minimize the amount of moisture in the air supply, and thereby minimize the amount of condensation in the dry pipe system piping.

Pressurized Nitrogen Supply

Some dry pipe systems require the use of nitrogen as a more economical alternative than an air compressor due to a lack of available and/or adequate electric power supply. More often, nitrogen is considered when the air supply will contain considerable moisture or where the temperature differential between the air supply and the dry system is quite pronounced, and relatively warm, moist air would be introduced into extremely cold piping. Nitrogen can also inhibit the accumulation of MIC (Microbiologically Influenced Corrosion).

Nitrogen cylinders are connected to the dry system similar to the air pressure supply arrangement, with an air pressure maintenance device to maintain required system pressure. The use of nitrogen cylinders introduces an additional maintenance item since the supply is limited, and therefore, the capacity and pressure of the cylinders must be monitored regularly. The small dry system leaks that go unnoticed with air supplied by an air compressor or shop air can present a very critical situation where the pressurized supply is limited. A reserve cylinder of nitrogen must be readily available at all times to replace a cylinder that has fallen below required capacity.

Dry Valve Intermediate Chamber

The majority of modern dry pipe valves have an intermediate chamber which surrounds the water supply face of the clapper, but when the system is in the valve closed position, this intermediate chamber is sealed off from water and compressed air supplies. The function of the intermediate chamber is to speed the action of the dry pipe valve as follows: Most dry pipe valves are designed so that the initial drop in air pressure when a sprinkler head operates will allow the clapper, or an auxiliary clapper, to raise slightly due to the supply water pressure. Some supply water is then forced into the intermediate chamber under the main clapper. Now the air/water pressure differential is destroyed because the water pressure is acting on a much larger area-the water supply pipe size surface plus the intermediate chamber surface. Consequently, the air pres-

sure is not capable of holding the clapper in the closed position.

Dry System Fire Department Pumper Connection

When the supply water pressure forces the main dry valve clapper to rotate on its hinge pin, it engages a latch mechanism which holds it in the open position. To reset this type of dry pipe valve, it is necessary to drain the system of water, unbolt and remove a face plate, manually release the latch, and free the clapper.

There is an interesting difference between the alarm valve and the dry valve which influences the design of a dry pipe system. In discussing the wet system, and specifically the fire department pumper connection to the system, the alarm valve served as a check valve to prevent the fire department from pumping back into the system supply line. Since the dry pipe valve clapper latches in the open position, the dry pipe valve cannot be used as a check valve, and a standard check valve must be installed in the main supply line.

Quick Opening Devices

Over time, effort has been made to overcome the head actuation and water discharge time delay encountered with a dry pipe system. This has been accomplished, to some degree, by the installation of quick opening devices, to either increase the rate of discharge of air from the system piping, or to accelerate the opening of the dry valve.

These devices, depending on their method of quick opening the dry pipe valve, are call exhausters and accelerators. Accelerators are not required soley on the basis of system volume. Accelerators are an option if the system exceeds a 500 gallon capacity and requires more than 60 s for water delivery at the sprinkler.

Accelerators

The movement of the diaphragm in an accelerator opens an auxiliary valve which admits compressed air to the intermediate chamber of the dry pipe valve. This immediately upsets the water-to-air pressure differential of the dry pipe valve. The air pressure on the air side of the clapper can no longer hold the clapper closed against the water pressure, and the additional air pressure in the intermediate chamber and the system water pressure forces the clapper open.

In order to allow the air pressure in the intermediate chamber to exert its full force on the underside of the clapper, the intermediate chamber drain must close automatically when the compressed air is introduced. Latched clapper or mechanical-type dry pipe valves do not have an intermediate chamber and, therefore, cannot be tripped through the use of an accelerator.

The operating principle of one such quick opening device is based on the use of two air chambers of equal pressure separated by a movable diaphragm. With this accelerator, one air chamber is designated as the inlet or lower chamber and is connected to the dry system so that the air pressure in this chamber is equal to the dry system air pressure at all times. The second chamber, called the upper or pressure chamber, is completely closed and sealed with the exception of a small, restricted orifice connecting the upper and lower chambers. This restricted orifice allows the pressure in the lower chamber to seep into the upper chamber so the pressure in the upper and lower chambers can become equal. The two chambers are separated by a diaphragm, and when a sprinkler head actuates, there is a sudden drop in the dry system air pressure, and at the same time, in the lower chamber. The restricted orifice is much too small to allow sufficient air to escape from the upper chamber to compensate for this sudden drop in air pressure in the lower chamber, and the upper chamber, or original system air pressure, exerts an unequal force on the diaphragm forcing it downward. This movement of the diaphragm actuates various mechanical mechanisms and valves, the end result being the immediate tripping of the dry pipe valve.

Today electronic accelerators are available. They have many advantages over mechanical devices. The simplicity and dependability of the electronic accelerator switches as a substitute for the mechanical accelerator are key. The following list is a summary of some of the key benefits of the electronic accelerator: (Golinveaux 2018)

- Facility Operator costs decrease due to less frequent maintenance issues.
- Fewer maintenance issues insure that the accelerators (critical to many systems) remain in service.
- Faster operation improves the performance of larger dry systems.
- With no moving parts in the pressure sensing area, maintenance and dependability are improved dramatically.
- Without a restricted orifice, false trips due to temperature changes are eliminated.
- Without a restricted orifice, maintenance and resetting issues related to anti-flooding devices and drain-back do not apply.
- Built-in low and high air pressure switches: small leakage will trigger supervisory alarm vs. the activation that might occur with a clogged restriction in a mechanical accelerator.
- Proven electronic technology used with electrically operated deluge and preaction systems.
- Consistent operating times improve water supply analysis and dry system performance.
- Accurate dry pipe valves trip times will assist in calculating dry pipe system performance.
- Insurance Risk assessment improvement due to dependability of service.

Exhausters

The movement of the diaphragm in an exhauster opens an auxiliary valve on the device which dumps, or exhausts, the system air pressure to the atmosphere. The auxiliary valve on an exhauster is of such a large size that great quantities of system air are exhausted very rapidly and, as a result, the system air pressure drops quickly to the dry pipe valve trip point. Once the dry pipe valve has tripped and water starts to fill the system, water is transmitted to the exhauster. The action of the water, combined with the mechanical design of the exhauster in a wet condition, closes the auxiliary valve automatically, preventing discharge of water to the atmosphere through the exhauster valve.

Dry Pipe System Drainage

Because a dry system is used in areas subject to freezing, it is imperative to keep the piping empty of water. Piping is pitched to facilitate (and guarantee) gravity draining of any water in the piping back to the main drain at the dry pipe valve when the system is drained. The required piping pitch is at least 1/2″ in every 10′ for the branch lines, and at least 1/4″ in every 10′ for cross mains and feed mains. In refrigerated areas protected with a dry pipe system, it is good practice to increase this pitch as much as possible. An increase ensures that water does not accumulate from condensation, and that all the water drains out quickly after the system has tripped and is being drained for resetting.

Pipe Support

An additional point to consider in pitching the piping on a dry system is the possibility that the construction supporting the sprinkler piping may settle under the additional weight of the sprinkler system. This can occur in light construction, especially that which was not originally designed to accommodate a sprinkler installation.

Refrigerated Areas

In refrigerated areas, and areas where the possibility of settling exists, the piping should be pitched at least 3/4–1″ in every 10′ for branch lines, and at least 1/2″ in 10′ for feed and cross mains. As mentioned previously, the branch lines, cross mains, and feed mains in these areas should be pitched as much as construction and floor height will permit using the 3/4–1″ in 10′ for branch lines and 1/2″ in 10′ for cross mains and feed mains as minimum criteria.

The main drain on a dry system has identical size requirements to the wet pipe system, with the drain size based on riser size.

Special Conditions

Due to construction conditions such as low rooms, or decks, a dry system, like the wet system, may have low trapped sections of piping that will not drain back automatically by gravity to the main drain at the dry pipe valve. These trapped sections on a dry system are obviously of greater concern than those on a wet system; trapped water in the piping of a dry system can freeze, expand, crack the piping, cause a loss of air pressure, trip the dry valve, and cause serious and expensive water damage. Another hazard is present when the dry system is shut down for repairs, which not only leaves the facility without fire protection, but also invites the possibility that a closed main control valve could be inadvertently left unopened, depriving the system of water supply for an unidentified period.

Maintenance Considerations

Maintenance for draining trapped areas on a dry system must be relatively easy and trouble-free. Maintenance personnel should readily drain these sections of piping after the system has tripped or be able to determine if water from condensation has built up in the low piping.

Drainage Capacities

Where a trapped section of piping contains five gallons of water or less, the auxiliary drain for these areas may consist of a valve with a brass plug or nipple and cap, to reduce leakage of system air through a defective drain valve. A trapped section on a wet system of similar capacity only requires a plug or nipple and cap. With the drain valve required for an auxiliary drain on a dry system, it is possible to carefully bleed off condensation water accumulation by barely cracking the valve without tripping it. This would be an impossible operation if a plug or cap had to be removed.

Drum-Drip Drain

Where the capacity of the low or trapped section of piping on a dry system contains in excess of five gallons, the auxiliary drain consists of a drum-drip arrangement. A trapped section of piping containing more than five gallons is large enough to allow condensation to collect, which could build up rather rapidly in the drain line. If not emptied frequently, this water could completely fill the drain line, freeze, and break a fitting. A drum-drip auxiliary drain is a collecting point for condensation water, an area which allows for a certain amount of expansion if the water should freeze. This provides for easy removal of a relatively large quantity of collected water without tripping the system.

Should the drum-drip fill, freeze, and crack the pipe or fitting in the condensation collecting nipple, the system would lose air and trip. Resetting the system back into service requires merely closing the upstream valve on the drum-drip, draining, and restoring the dry system to service, and then replacing the drum-drip assembly at the convenience of the maintenance department. No prolonged shutdown of the system is required.

A drum-drip auxiliary drain consists of two 1″ valves, one on the system supply side of a 2″ diameter by 12″ long pipe, (or unit of equivalent volume), which is the condensation collector, and one on the discharge side of the condensation collector.

The valve on the supply side is kept open while the valve on the discharge side is closed and sealed with a brass plug or nipple and cap to prevent air from escaping should a valve leak. When condensation collects in the piping of a trapped section, it drains by gravity through to the 1″ drain line, through the open 1″ supply side valve, and collects in the condensation chamber. During regular maintenance, the supply side valve is closed, the discharge valve opened, and the plug or nipple and cap removed to drain out water that has accumulated.

Low Points

If several low points or trapped areas on a dry system are all located in the same general area, it will facilitate maintenance to install a tie-in drain. Connecting or tyeing in all the low point drains through a common 1″ pipe, and providing one auxiliary drum-drip drain at the termination of this tie-in drain, increases the reliability of all the low points being drained.

Dry System Design Features

This section will only deal with a few of the basic design considerations which are required when a dry system is installed.

Calculating the Area of Application

Depending on the design of the dry system, and not including unusually large supply pipe or long runs of supply pipe, the capacity of a dry system varies between 3/4 and 1–3/4 gallons per sprinkler. For an existing system, a good rule of thumb in estimating (not engineering) the capacity of a dry pipe system is to multiply the number of sprinklers by 1.25 gallons per head, and add the capacity of the bulk, or supply, piping if it exceeds 20 in length. This 1.25 gallons per head includes the line piping and the cross main piping when the dry pipe system layout is relatively simple.

When calculating the accurate capacity of a dry pipe system, the pipe of different sizes must be tabulated, totaled, multiplied by the number of gallons per foot capacity of each pipe size.

Qualification test approval of an excess capacity dry pipe system cannot be based entirely on the 60-s water delivery. As with all fire protection systems, the evaluation of all the elements involved must be considered by a qualified individual. In this case, the 60-s water delivery is merely one facet of an acceptable dry pile system. The occupancy being protected, construction material of the structure, floor heights, even the availability of the public fire department, their equipment, and their manpower, must be considered and evaluated.

Location of Dry Valves

The location of the dry valves is an important element in the design and function of the system. The dry valve or valves should be located as close as possible to the system to keep the amount of supply piping to a minimum and thereby reduce the total pressurized air capacity of the system.

The piping up to the clapper of the dry valve is filled with water, and in some dry valves there is a priming water seal on the dry side of the clapper in the dry valve. Naturally, this wet portion of the system must be kept from freezing. This can be done by locating the dry valve or valves in a warm area of the building or locating them in a special dry pipe valve enclosure to protect the wet section of the dry system from freezing.

A dry pipe valve house may be constructed either inside or outside of the building being protected by the dry system. The valve house must be of insulated construction with walk-in refrigerator-type door and hardware so that a small capacity thermostatically controlled heater can maintain at least a 40 °F temperature inside the valve house, regardless of the outside temperature. The heater may be electric, steam, or any other practical, reliable, and economical heat source.

To increase the reliability of maintaining adequate heat in the valve house or enclosure, a low temperature alarm device can be provided. Should the heat source fail to provide 40 °F, the low temperature alarm will activate and transmit an alarm signal. As with the low air system on the dry pipe system, the low temperature trouble alarm signal should be transmitted to a constantly attended location for immediate attention.

Immediate attention is not quite as critical as it is with the receipt of a low air trouble alarm, because the insulated valve house or enclosure will retain sufficient heat to prevent a hard freeze for a short period of time, but attention to the problem must be given high priority. In addition

to heat, the valve house must have adequate lighting to facilitate maintenance and inspection of the dry pipe valves and be of a size that will provide easy access to all sides of the dry pipe valve and dry valve trim.

Water Columning

Dry system risers leaving the valve house or enclosure have been known to fill with water and freeze due to water columning. This is due to the accumulation of water in the riser resulting from leakage past the water seat of the dry pipe valve clapper or slow drainage of condensation back to the dry valve. These risers are usually large diameter pipes and when the water freezes there is enough capacity to absorb the expansion. However, the ice in the pipe can render the system inoperative by preventing water passage in the pipe, or transporting the ice to small piping or sprinkler head orifices and blocking them. The accumulation of water in the riser may not rise to a height in the riser past the roof or wall of the heated valve house and into the freezing area. However, due to the operational characteristics of the ordinary dry pipe valve, a relative minor accumulation of water in the riser can produce additional weight on the air side of the clapper and increase the trip point of the valve. When the system operates, and the system air pressure is released, the valve will not trip as quickly because of the weight of water on the dry side of the clapper.

The dry pipe valves equipped with intermediate chambers overcome water columning. The automatic drain feature, which keeps the intermediate chamber open to the atmosphere, will not allow any water accumulation. This does not prevent slow drain back of condensation from the system from building up in the riser. In addition, there are dry pipe valves that do not have intermediate chambers, so each dry system design where water columning could present a problem should include an approved method of eliminating this water accumulation. This may be accomplished by installing a high water level signaling device, a float – type automatic drain valve, or an automatic drain device.

Low differential dry valves, because of their design function, are not subject to the weight of water on the clapper causing a time delay in operating. The possibility of the water accumulating and freezing in the riser makes it necessary to install a high water level signaling device or an automatic drain even though the dry valve is the low differential type.

One design feature which can make a dry pipe system more expensive than a wet system is the necessity for concealing the piping above a ceiling and installing the sprinkler heads below the ceiling in a pendent position, or when construction or head room mandates turning the heads in the pendent position on exposed piping. Only special dry pendent sprinklers may be used in the pendent position on a dry system. These special heads are described in detail in the chapter on Automatic Sprinkler Heads, but basically, they are constructed to prevent water from collecting in the head or in the drop nipple to the pendent sprinkler head. When the system trips, water would naturally collect in the shaft of a head-turned pendent or in the drop nipple from concealed piping to the pendent sprinkler below the ceiling. This water would not drain off when the system was drained, and the dry pendent sprinkler prevents water from entering the pendent head shaft or drop nipple.

Deluge Systems

Although sprinkler systems have an admirable performance record of property protection, some hazards require special protection in addition to the building protection provided by the sprinklers. This is evidenced by the relatively small number of serious fire losses that have occurred where combustion heat, intensity, and spread overpowered the sprinkler system. Areas where easily ignitable and fast burning material exist require special hazard systems.

One special hazard system to be considered is the deluge system (See Fig. 9.3). The wet and dry pipe systems previously described have closed sprinkler heads, heads with fusible links, or fusible elements that only discharge when the fus-

ible link or element is heated to a predetermined temperature.

A deluge system consists of open sprinkler heads. The sprinkler heads on a deluge system are similar to the heads on a wet or dry pipe system, but they do not have any fusible links or elements. When the system is activated, water is discharged from every head on the system, simultaneously creating a deluge of water.

When the fixed temperature release (1) is activated by fire, pressure in the release system escapes from the open device, allowing the pneumatic actuator (2) to open. This release pressure from the priming chamber (3) of the deluge valve, allowing the valve to open. Water flows into the system piping and to the alarm devices, causing the pressure switch (4) to activate an electric alarm and/or operating a mechanical water motor alarm. Water flows from all open sprinklers or nozzles (5). When the deluge valve operates, pressure closes the PSOV (6) cutting off the water supply to the priming chamber, latching the deluge valve in the open position.

The piping on a deluge system is obviously empty because all the sprinkler heads are open. The deluge valve prevents the water from entering the system piping until a signal from the detection system trips the deluge valve. A deluge system valve is opened to admit water to the system by the action of a separate detection system. This detection system may be electric, pneumatic, or hydraulic. The deluge system is designed to wet down the entire area protected by the system. One typical example is the protection of a cooling tower.

For example:

By virtue of its design, there are numerous wind currents surging through a cooling tower, and this is especially true when it is not in operation and dry. With a wet or dry pipe sprinkler system, the heat from a fire in the northwest corner could easily be deflected by wind currents and open closed heads in the southeast corner. Obviously, the discharge from these fused heads would have no suppression effect on the fire origin. With the deluge system valve activated by the detection system in the cooling tower, no matter which detector is activated (in this example, the heat would activate the detector in the southeast corner), the deluge valve will be tripped and water discharged over the entire area, extinguishing the combustion in the northwest corner.

Fig. 9.3 Deluge
System. (Courtesy of
Viking Corporation)

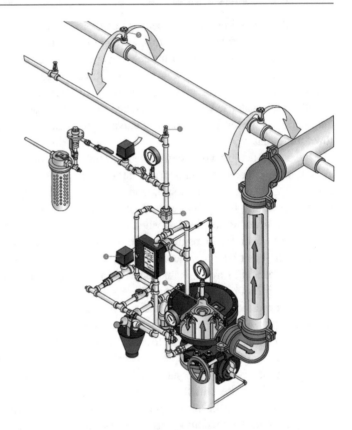

Fig. 9.3 Deluge
System. (Courtesy of
Viking Corporation)

Another advantage of a deluge system is its cooling capacity. The deluge of water not only controls and extinguishes the combustion, but also cools all equipment and structural members within the area of the system discharge. This prevents damage of equipment by radiant heat, or structural steel deformation by combustion heat possibly resulting in severe building structural damage.

The first deluge valve to consider is a quick opening valve having only one moving part, the clapper, which is kept closed by water pressure. This valve can be activated by a manual release or a hydraulic, pneumatic, or electric detection system. The deluge valve has three chambers: the top chamber which is under water pressure, the discharge outlet which is normally dry, and the inlet chamber or incoming water supply. The chambers are separated by the valve clapper. The water pressure on the top of the valve clapper, or top chamber water pressure, is the same as the inlet chamber or water supply pressure. This is accomplished by a small priming line which

takes supply from the water supply and bypasses the clapper, discharging into the upper chamber. This priming line is referred to as the bypass line.

The priming line has a control valve, a restricted orifice to limit the flow of water through the line when the system is activated, and a strainer to prevent the passage of small particles of sand or dirt which could clog the restricted orifice.

The top of the valve clapper, which forms the bottom of the top chamber, is twice as large in area as the bottom of the clapper where it covers the top of the water supply inlet chamber – a 2:1 area differential design.

When the valve is in the closed position, the supply water pressure is equal on both sides of the valve clapper. On the inlet side, this water pressure is acting on approximately 28 sq. in. for a 6″ water supply to the valve, while the same water pressure is acting on approximately 56 sq. in. on the top of the clapper. With a supply pressure of 75 psi (pounds per square inch), approximately 2100 pounds of supply water pressure pushing up on the bottom of the valve clapper is opposed by approximately 4200 pounds pushing down on the top of the valve clapper. (These numbers are used for example only, and are therefore approximations.)

When the detection system is activated, the water pressure on the top of the clapper is relieved through the release line, and the clapper is lifted by the supply water pressure underneath. Water fills the piping and is discharged from all the sprinkler heads on the system. The priming, or bypass, line restricted orifice cannot make up water pressure fast enough to overcome the released pressure on the top of the clapper when the detection system has opened the release line and relieved the pressure on the top of the clapper.

Most other deluge valves have a clapper that is latched in the closed position. One such valve has a diaphragm with equal pressure on both sides. Activation of the detection system relieves the pressure on one side of the diaphragm and, when the diaphragm moves, it operates a shaft and lever mechanism which unlatches the clapper.

Another manufacturer provides a deluge valve that has a weight held in a fixed position. When the detection system is activated, the weight is released and drops, opening the clapper latch. The water pressure rotates the clapper to the open position.

The water supply to the deluge system is controlled by a valve in the water supply line to the system. As with the dry pipe system, once the system has tripped and the clapper has rotated to the open position, any water pumped through the fire department connection will pass back through the deluge valve into the supply, if the fire department is pumping water in at a higher pressure than the system supply pressure. Therefore, it is necessary to install a check valve in the water supply to the deluge system below the deluge valve to prevent the fire department from pumping back into the water supply. The fire department pumper line will be provided with a check valve to prevent water from leaking out of the siamese connection, or from discharging out of an uncapped siamese, when the deluge system is activated and water fills the system.

When the deluge valve is tripped and the system piping fills with water, the small alarm line is also filled with the pressurized water. This alarm line can supply a pressure switch which will make an electrical contact when the pressurized water enters the device, and can also operate a water motor alarm.

Since the clapper on deluge valves is held closed by a mechanical latch or water pressure, a retard chamber is not necessary. It is used in other systems to prevent supply water pressure surges from activating the alarm devices. Another device that cannot be used is the paddle-type water flow switch. Because the water rushes into the system piping when the deluge valve is tripped, the paddle in the pipe would either be ripped off or severely damaged by the velocity of a sudden rush of water. (This condition also exists on the dry pipe system.) Paddle-type water flow switches can only be used on wet pipe systems where the piping is constantly filled with water. When the system operates, the water surrounding the paddle in the piping merely deflects with the movement of the water.

Pre-action Systems

The dry pipe system was an adaptation of the basic wet pipe system; the pre-action system is a sophistication of the dry pipe system. The pre-action system provides more efficient fire protection than the dry pipe system and minimizes the possibility of water damage-both accidental and fire related-but these features introduce additional cost.

When installing fire protection in a freezing environment, it is necessary to accept the added cost of the dry pipe system, both in equipment and labor of installation. Further, the additional cost of the pre-action system must be compared with the protection value for both fire and water damage that the pre-action system provides. A pre-action system, except for the pre-action valve, resembles a dry pipe system, with the same piping arrangement, supports, sprinkler head coverage for the hazard protected, and closed head sprinkler heads. However, the pre-action system has a second system: a detection system.

The piping for a pre-action system is dry, but it cannot be referred to as a dry pipe system. Dry pipe systems contain piping with pressurized air

Fig. 9.4 Pre-action
System (Courtesy of
Viking Corporation)

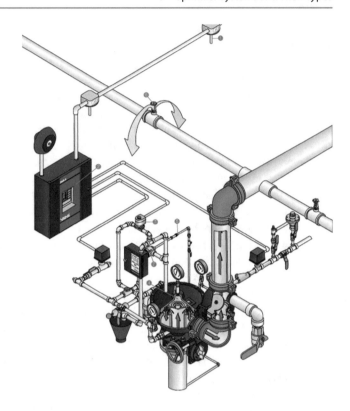

or nitrogen and has a valve that is tripped by the
loss of pressurized air or nitrogen in the piping. A
basic pre-action system has a water supply that is
held back by a pre-action valve. The fusing of a
sprinkler head on the system will not activate the
pre-action valve and admit water to the system. A
piece of piping on the pre-action system can be
removed and the pre-action valve will not trip,
nor will water be discharged. The only action that
will trip a pre-action valve and admit water to the
system piping is the activation of the detection
system. This statement is not completely true
because there are manual devices available to trip
the pre-action valve, but the only automatic way
to trip the pre-action valve is activation of the
detection system.

The Pre-Action Valve

Water supply to the system is held back by the pre-
action valve, shown in Fig. 9.4. When the detection
system is activated: (1) When the detector (1) is
activated by fire, a signal is sent to the Release

Control Panel (2). The panel sends appropriate
alarm and trouble signals and, at the same time,
signals the release of the solenoid valve (3). The
deluge valve priming chamber (4) is then vented
faster than water is supplied through the restricted
orifice (5), allowing the deluge to open. The
water enters the system piping, but until a sprinkler
(6) activates no water is discharged. When the deluge valve operates, pressure closes the valve (7)
cutting off the supply of water to the priming chamber, latching the deluge valve in the open position.
At this point, no water is discharged until the heat
from the combustion fuses a sprinkler head.

Detection System

A sprinkler head, even those temperature-rated at
135 or 165 °F, are not considered to be sensitive
detectors. When the heat from combustion has
risen sufficiently to fuse a sprinkler head, considerable damage can be inflicted on the occupancy
before the sprinkler discharge suppresses the
combustion and cools the area.

Fig. 9.5 Water mist conductivity testing. (Photograph by Robert Till)

By using a detection system that is much more sensitive than a sprinkler head to combustion in the very early stages, a fire alarm will be activated and personnel may be able to extinguish the combustion before sufficient heat develops to fuse a sprinkler head. The detection system that has been activated not only actuates the alarm system, but also trips the pre-action valve and admits water to the pre-action system. No water has been discharged at this point because none of the heads have fused, but the pre-action system has become a wet pipe system. If the occupants cannot extinguish the combustion with hand extinguishers or hose, and the combustion continues to develop, sprinkler heads will fuse, water will be discharged from the open heads, and the combustion controlled or extinguished.

A pre-action system provides an early warning of a fire condition by the activation of the detection system in most cases long before there is sufficient heat to fuse a sprinkler head.

The detailed description of various detectors and detection systems will be found in the Chap. 3. In order to explain more fully the advantages of a pre-action system, two detection systems will be briefly described in this section.

A smoke detector system that will identify the products of combustion produced while the combustion is in the incipient stage-that is before there is visible smoke, noticeable heat, or odor.

A rate-of-rise detection system uses detectors that are activated when the temperature rises in excess of 15 °F per minute, regardless of the ambient temperature of the area. When the detection system is rate-of-rise, the pneumatic tubing also maintains a small air pressure, and should a leak occur and the tubing pressure be released, two actions will take place simultaneously. First, a trouble alarm will be activated, and secondly, the pre-action valve will automatically be tripped admitting water to the piping.

Either of these sensitive detection systems will, in most cases, activate long before there is sufficient heat to fuse a sprinkler head or damage equipment. Should a fire in the area get out of control, the pre-action piping has been primed, full of water, and ready to discharge through the first sprinkler head that fuses. This early alarm warning will allow occupants alerted by the alarm to investigate and determine the source of the combustion, and take action to extinguish the combustion, shut off power to a smoldering wire or panel board, etc., before the combustion can develop enough heat to fuse a sprinkler head.

Protection of Equipment

When sprinklers must protect an area containing computer or electronic equipment, a pre-action system is recommended, even though water damage from a non-fire situation such as a broken sprinkler head or a leaking fitting may be highly unlikely. Many objections are raised when an insurance carrier requires that a computer room be protected with automatic sprinklers. There are several good reasons for this requirement.

A computer room is situated in a building that is totally protected with automatic sprinklers, but there is no sprinkler protection in the computer room. Should a fire in the computer room get out of control, involve the entire room, and then break out into the sprinklered area of the building, it would, in all probability, overpower the sprinkler system, and involve the entire facility. When objections are raised to protecting the computer room or the electronics room with sprinklers, it can be stated that when a fire develops enough heat to fuse a sprinkler head, there is sufficient heat to damage or destroy all of the equipment in the room. When a fire reaches this condition, water damage is the least of all concerns.

Supervising the Piping System with Air

An additional safeguard against mechanical, non-fire, discharge of water with a pre-action system is to supervise the pre-action system piping with compressed air or nitrogen. In fact, the NFPA requires that pre-action system piping shall be supervised if the system serves more than 20 heads. In such cases, a small air pressure is maintained in the pre-action system piping, and this air pressure is monitored by a low air pressure alarm device. Should there be a small loss of air pressure through a defective fitting or pipe crack, the air pressure will drop to a predetermined point and activate the low air alarm. No water will fill the piping when the low air alarm is activated because, as previously explained, only the detection system can automatically trip the pre-action valve and admit water to the system.

The advantage of supervising the pre-action system piping is to prevent any leaks when the piping is filled with water. This could occur when the detection system has been activated and the piping has filled with water, but the combustion has been extinguished with an extinguisher before sufficient heat had developed to fuse a sprinkler head.

If the detection system is electrical, it is supervised to protect against a wiring break, short circuit, or ground. When one of these faults occurs, a trouble alarm is activated, and the pre-action valve tripped, admitting water to the piping. Supervision of the detection system is necessary because the detection system controls the sprinklers. When the detection system is rendered inoperative, the pre-action valve is automatically tripped, water fills the piping, and sprinkler protection is maintained until the fault can be corrected and the total pre-action system restored to operational status.

Use of Preaction Systems

Since the piping of a pre-action system is dry, these systems can be used in areas subject to freezing. The same design criteria is used for a pre-action system in these areas: When heads are installed in the pendent position they must be dry pendent heads. When a pre-action system is used in areas not subject to freezing, regular pendent heads may be used, but only if their use is approved by the insurance interest. Pre-action systems are not recommended for use in cold storage areas such as freezers or walk-in refrigerators because of the time element between detection, subsequent filling of the piping with water, personnel response, and possible extinguishment of the combustion before a head fuses.

Most early warning detection systems are not affected by low temperature. A rapid rise in temperature activating a rate-of-rise detection system, or the products of combustion with a detection system, could fill the piping with water long before any action was taken to drain the pre-action sprinkler piping. This water in the piping with no movement, and the piping at the temperature of the freezer, would allow the water to freeze very quickly, which would present a difficult and expensive problem. The freezer temperature could not be lowered to below freezing long enough to allow the frozen water in the piping to turn back to water and be drained off. With the piping completely clogged with ice, the sprinkler system is useless. The only alternative is to remove the piping, thaw the ice, and reinstall the piping system. During this expensive and time-consuming operation, the one fire protection advantage would be to keep the detection system in service. The use of pre-action systems in areas subject to freezing is a matter of engineering judgment combined with a thorough familiarity with all the conditions surrounding the installation. The following is an actual case example where the human element was not considered.

A pre-action system was installed to protect the coal conveyor of a steam power generating facility. A few years after installation, which was also a few years after operating instructions had been received and reviewed with the operating personnel (and several personnel changes had occurred) a bearing on a conveyor roller overheated and ignited the accumulation of lubricating grease. It was a small fire, but intense enough to activate the detection system, which opened the pre-action valve admitting water to the piping and transmitted an alarm to the control room. Personnel were dispatched to the area and the fire was quickly extinguished with a dry chemical extinguisher. Since the heat from the fire diffused by the extreme draft conditions often found in conveyor enclosures was not intensely concentrated enough to fuse a sprinkler head, no water was discharged. The sprinkler system looked no different after the fire than it had looked before the fire, and no different than it had looked to the operating personnel every day over the years. No one realized, or was aware of the fact that, the system had tripped and should have been drained down and the valve reset. The result was obvious, the mechanical repair to the conveyor was completed in short order and the personnel returned to their normal duties. The piping froze overnight, and when the sun shone down on the metal roof of the conveyor enclosure the next day it thawed

some of the piping and water began to gush from a fitting that had cracked under the pressure of the frozen water. Considerable water damage occurred before the leaks were discovered, and a considerable expense was involved in repairing the sprinkler piping. The human element must be a serious consideration in the judgment to use a preaction system under circumstances where the reliance on adequate and informed operating personnel is a serious consideration.

Pre-action System Design Features

As stated previously, a pre-action system has the same three essentials as the wet and dry pipe systems: alarm, water supply valve control, and fire department pumper connection.

The difference is the pre-action valve, a deluge valve, or a *pre-action deluge valve*. The deluge valve is not activated by a flow of water as is the alarm valve, or by a loss of pressurized air or nitrogen as is the dry pipe valve. Instead, deluge valves are activated or tripped by a signal from a detection system. A deluge valve is latched or held shut by water pressure and, regardless of what may affect the system piping or sprinkler heads, the valve does not trip and discharge water to the system until a detection system releases the valve to allow water flow. Once a deluge valve is tripped and water flows, water can enter an alarm line and activate a pressure switch, close electrical contacts, and activate an alarm system. In addition, water can also be admitted to the alarm line to operate a water motor alarm. The pressure switch and water motor alarm are identical to those used on the wet and dry pipe systems.

A word of caution The use of a *paddle-type flow switch* is not allowed on a pre-action system for the same reason this device could not be used on a dry pipe system. When the deluge valve opens, there is a rush of water into the system, and the velocity of this volume of water would seriously damage the paddle of the flow switch in the pipe. *Do not use flow switches on dry pipe or pre-action systems.*

Another important design feature of pre-action systems that have supervised piping is maintenance of a constant air pressure. A rubber-seated check valve is necessary to maintain air pressure in the system riser, and is installed on the system side of the pre-action valve. Rubber seated check valves are used so that the clapper of the check valve will form an air tight seal. With metal-to-metal check valve seats, there can be a tiny metal defect that would allow air pressure to escape, or a grain of sand or dirt can become lodged under the clapper and destroy an air seal.

The fire department pumper connection is connected into the pre-action system riser on the system side of the deluge valve between the deluge valve and the rubber-seated check valve. This connection is made at this point on the system so the fire department could pump into the system even if the deluge valve had not tripped and was in the closed position.

As with the dry pipe valve, the deluge valve does not act as a check valve to prevent the fire department from pumping back into the supply. It is necessary to have a check valve in the supply riser downstream of the deluge valve.

If the water supply control valve is to be installed in the pre-action riser, it is customarily installed below the deluge valve and above the water supply check valve.

There is always a check valve in the fire department pumper line to prevent water in the riser from discharging into the siamese connection when the deluge valve trips. This check in the pumper line is required regardless of the type of system it serves. A small, usually automatic ball check valve is installed on the siamese side of the check valve at the low point to automatically drain off any condensation water or leakage past the check valve to prevent freezing in the siamese.

Types of Pre-action Systems

Basic Pre-action System: Single Interlock

What has been described is the basic pre-action system, and a single interlock pre-action system.

A detection system automatically activates a pre-action valve and water is admitted to the system piping, but no water is discharged until a sprinkler head is activated. The fusing of a sprinkler head independent of the detection system will not initiate the filling of the piping with water, and therefore no water will be discharged from the activated head until the detection system is activated and trips the pre-action valve.

System piping may be supervised with pressurized air or nitrogen to detect leaks, broken piping, or a defective or fused sprinkler head before the pre-action valve is tripped and water enters the system. A low air trouble alarm will indicate the loss of the supervision pressurized air or nitrogen. Only activation of the detection system, manual operation of the emergency release, or malfunction of the detection system will admit water to the system piping.

Double Interlock

The double protection pre-action system is used in areas where maximum protection against accidental filling of the piping with water is required-such as refrigerated storage areas, sensitive electronic equipment areas, or computer rooms. This is referred to as a double interlock pre-action system. With the double interlock pre-action system, both the detection system and a sprinkler head must be activated before the water will enter the system piping. Activation of either the detection system or a sprinkler head fusing will only activate an alarm.

The piping for this system contains pressurized air or nitrogen similar to the dry pipe system. Its design utilizes a deluge valve (pre-action valve) which can only be opened when pressure is reduced in the sprinkler piping and the detection system operates.

This system must only be used with the approval of the insurance interests having jurisdiction and designed by experienced fire protection engineers because it greatly increases the non-fire water damage reliability and the potential for delay in applying fire suppression water discharge.

Non-interlock

The combined dry-pipe and pre-action system can be activated by operation of the detection system or the fusing of a sprinkler head. This is the regular non-interlock pre-action system. One major design feature of this system is the use of a dry pipe valve in lieu of the deluge valve, but it differs from the basic dry pipe system because of two factors: Both the release of pressurized air or nitrogen in the piping, and the activation of the detection system will trip the dry pipe valve.

One major advantage of this system is that, should the detection system fail to operate or malfunction, the system can operate as a dry pipe system. By increasing the fire protection reliability, this system decreases the non-fire water damage potential. Should the sprinkler piping or a sprinkler head by damaged, the pressurized air or nitrogen will be released, the dry valve will trip, piping will fill with water, and the water will be discharged.

The advantage of this system over a basic dry pipe system is the detection system sensitivity which will provide an early warning alarm. Activation of the detection system will trip the dry valve and fill the piping with water, but water will not discharge until a sprinkler head fuses.

As previously described, the basic pre-action system valve must receive a signal from the detection system to allow water to flow in the piping, but with this system either the detection system or the fusing of a sprinkler head will activate the system.

Quoting from the National Fire Protection Association Standard for Installation of Sprinkler System (2016) Sect. 7.4.2.2:

> Combined automatic dry pipe and pre-action systems shall be so constructed that failure of the dry pipe system of automatic sprinklers shall not prevent the detection system from properly functioning as an automatic fire alarm system.

Water Spray System

In discussing deluge systems, there are numerous instances requiring water discharge over an irregular mass, like a transformer, or a discharge of very fine water droplets in order to provide more cooling or to afford maximum cooling on a surface exposed to the radiant heat of a fire. The discharge pattern of a sprinkler head may be too uniform, and the water droplets too large to accomplish this protection, so the sprinkler heads are replaced with water spray or nozzles (water fog nozzles). Although water fog is a trade name, it is sometimes used to describe water spray systems, regardless of the manufacturer. Since the current sprinkler heads are referred to as spray sprinklers, the term water fog has gained even greater general use to identify true water spray systems.

A water spray system is a deluge system using specially designed nozzles in lieu of open sprinkler heads on a deluge system. Being a deluge system, the deluge valve is activated by a detection system. When the deluge valve is activated by a signal from the detection system, water is discharged from all nozzles on the system simultaneously. Unlike the requirements for sprinkler heads, water spray nozzles can be spaced, located, and positioned in any configuration that provides effective coverage of the protected area. The water spray pattern, velocity, and density determine the location and positioning of the water spray nozzles.

Determination of the proper spray pattern and the positioning of the nozzles to take full advantage of the spray pattern requires experience, expertise, and professional judgment. Each water spray system is individually designed, tailor made, for protection of a specific hazard. NFPA publishes a Code (NFPA 15) that furnishes technical requirements for the design and engineering of water spray systems based upon sound engineering principles, test data, and field experience. But the selection of nozzle patterns, velocity of discharge, density, positioning of nozzles, etc., is the responsibility of an experienced professional.

Some hazards to be protected with water spray are so unique that a fire protection company will construct a mock-up of the equipment, using the actual flammable liquid involved at their testing site. The test data obtained is then used to design and engineer the water spray system for the hazard. Combustion models are also being calibrated to be used in system design.

There is a very thin line separating water spray and standard sprinkler protection: A water spray nozzle and sprinkler head discharge differs in the discharge pattern, velocity, particle size, and density. Water spray systems may be installed to satisfy several fire protection situations; extinguishment, controlled burning (limitation of fire spread), exposure protection, and fire prevention. Although a water spray system may be designed to perform all of these functions, they will be evaluated individually.

Extinguishment

In the history of sprinkler head development, one of the primary fire extinguishing advantages of the spray sprinkler head over the old-style sprinkler head was its ability to break up the water discharge into smaller particles for better cooling and smothering action when the droplets were converted to steam. Water spray system nozzles produce a water spray consisting of very small water particles that are easily converted to steam. This action not only takes heat from the combustion in changing the liquid to vapor, thus cooling the combustion, but also the steam produced keeps oxygen from reaching the fire and produces a positive smothering action.

Water spray systems have been used, to a limited degree, to extinguish certain flammable liquid fires by emulsification, dilution, or both. Emulsification is the mixing or combining of the flammable liquid and water to form an emulsion: In other words, a state of suspension of water droplets in the flammable liquid, similar to the fat content in milk. With emulsion extinguishment of oil fires, the flames in the vapor-air mixture above the surface of the burning liquid are diluted by the water vapor of the water spray to a point where there is no longer a flammable vapor-air mixture and the fire is extinguished.

Some water spray droplets are vaporized as they pass through the heat of the flames, but the primary source of effective water vapor comes from the water film in the emulsion at the sur-

face of the burning liquid. The film is vaporized by the heat of the burning liquid, and this vaporization process reduces the temperature of the burning liquid. The vaporization of water in the emulsion also reduces the volume of flammable vapors and produces a barrier on the surface which excludes air (oxygen) from the combustion.

Vapor dilution, the water film in emulsion at the surface of the burning liquid, can only be accomplished if droplets in the water spray are large enough to pass through the flame area without being vaporized and reach the surface of the liquid. The flammable oil vapors immediately above the surface are too rich to burn until they rise and mix with air to form a combustible mixture. The vaporizing emulsion of water and flammable liquid at the surface in this rich vapor area produces a mixture of water and oil vapor that will not support combustion, and the fire is extinguished.

Lighter liquids, like kerosene, do not form an emulsion with sufficient stability to cause extinguishment by this method. To extinguish these lighter liquids and oils, the vaporization of the water spray droplets as they pass through the flames produces a cooling and smothering extinguishment.

Dilution is applying sufficient water to dilute the flammable liquid to the point where it cannot produce sufficient flammable vapors to support combustion. Dilution of a flammable liquid usually requires so much water that it becomes an impractical fire protection consideration.

Controlled Burning

The deluge concept of a water spray system, coupled with the cooling action of the water spray, can effectively limit the spread of a fire, even if the fire is of such intensity that the water spray system cannot complete extinguishment. In some specific instances a burning material cannot be extinguished regardless of the amount of water applied, and the intent of the water spray system is merely to confine the combustion to the point of origin. Another function of controlled burning is confinement of a fire that should not be extinguished, for example, a fire at a gas line leak. As long as the fire continues to burn, the escaping gas is consumed. But if the fire were extinguished, the escaping gas could spread and create a life safety situation or collect in a low point some distance from the leak, make contact with an ignition source, and ignite or explode. The water spray system will be designed to keep the surrounding piping and equipment cooled to prevent explosions, fires, or damage from the radiant heat. The water spray discharge will also keep the fire controlled to reduce the intensity of the radiant heat produced.

Exposure Protection

Many water spray systems are designed specifically to remove or reduce the heat from the surface of structures or equipment exposed to the radiant heat of a fire. A research firm conducted tests to determine the effectiveness of deluge systems with standard sprinkler heads on aircraft jet fuel spill fires. The data from these tests accents the value of water spray structural steel exposure protection. For example:

> One test consisted of a fuel spill of 900 to 1300 sq. ft and involving several hundred gallons of jet fuel. The very turbulent fire plumes that were formed by this fire attained velocities in excess of 50 miles per hour and steel specimens, 6 × 8 I-beams located at a height of 40′ above the fire, and at the ceiling 60' above the fire, attained temperatures between 800 and 1200 °F within a very few minutes. At such temperature levels, structural steel under a load condition will deform. When steel deforms, severe structural damage can occur, therefore, the exposure protection of a water spray system can easily be justified when a steel structure or structural steel supports are exposed to an intense fire condition.

Designs have been prepared using the water spray system as a *water curtain* separating the fire area and the equipment to be protected. Water spray curtains can be effective under certain conditions, but direct application of the water spray is much more efficient exposure protection.

Fire Protection

An example best illustrates this function of a water spray system:

> A bank of three large oil-cooled transformers at a power generating facility each contain 10,000 gallons of oil. An oil reservoir on one transformer is ruptured and oil starts to spread under the other transformers. Every hot surface or high voltage electric contact presents an ignition source. The water spray systems surrounding each transformer are actuated by tripping the water spray deluge valves automatically by a detection system. The water spray will disperse the oil and, to a degree, dilute and thin the oil spill, which very effectively keeps all surfaces cool, preventing ignition. Should one transformer become ignited, water spray on the uninvolved transformers will prevent or control radiant heat damage.

Using transformers as an example serves two purposes: (1) to point out the use of water spray systems as a fire prevention system, and (2) the protection of oil-filled transformers with water spray. An electrical short in a high voltage transformer can create enough heat to vaporize the oil. The resultant pressure can rupture the oil reservoir casing, spewing hot oil over hot transformer surfaces. The resultant fire is extremely intense, and exposure protection with water spray of the other uninvolved transformers becomes of the utmost importance.

Electrical Conductivity

Water spray nozzles placed at a predetermined distance from electrical equipment will not conduct an electrical current on the water droplet discharge. This is a major consideration in fighting a transformer fire, or a fire in any high voltage electrical equipment.

Water Spray Versus Sprinkler Systems

Before examining the design features of water spray systems, it is important to clarify the interface between sprinkler and water spray systems. Water spray systems are not intended to replace automatic sprinkler systems, although water spray systems may, under certain circumstances, supplement or compliment sprinkler systems in protecting special hazards.

Design Considerations

The first determination in recommending a water spray system are

1. the nature of the hazard.
2. the specific purpose for using this system.

The hazards most frequently protected with water spray systems are those involving flammable liquids, gas tankage, associated piping, devices, and equipment, electrical equipment such as transformers and oil switches, generators, and other rotating electrical machinery. Primary design considerations for a water spray system are the anticipated heat produced by the combustion, and the radiant heat from exposure fire. Once this is determined, a density (gpm per square foot) sufficient to absorb the combustion and exposure head can be calculated.

Nozzle Selection

Another consideration is the droplet size produced by the nozzles to determine if the water spray should be coarse or fine. The extinguishment of a fire occurs at an interface between the surface of the burning material and the base of the flame, so it is necessary to use a nozzle that will not only produce a water spray at the required density, but also a water spray delivered at a velocity and mass (droplet size) to penetrate the heated air currents and reach the fire zone. In the case of exposure protection, the surface of the equipment must be completely covered in spite of heat waves radiating off the metal surface.

Nozzles for water spray systems come in all sizes and shapes with an equal multitude of discharge patterns and methods of developing the spray pattern. The engineer must select the proper nozzle to fit the conditions: the nature of the hazard, the purpose of the system, and in severe wind or draft conditions, the deflection of the water spray. Generally, high velocity nozzles produce a cone pattern, and low velocity nozzles deliver a spheroid or cone pattern consisting of a much finer spray. With most water spray nozzles, the higher the velocity, the coarser the spray or water droplets, and the greater the range of reach of the spray. On the market there are several types of very high velocity, high capacity, spray nozzles designed primarily for transformer protection. With heavy capacity few of these nozzles are required. The transformer can be removed without disturbing the water spray system piping because the nozzles can be located at some distance from the transformer, with wide spacing between nozzles. Water spray nozzles produce their characteristic spray pattern in different ways depending on the manufacturer. One spray nozzle develops a high rotary motion of the water inside the nozzle head by the use of spiral passages, another by impingement of small, carefully directed jets of water through opposite multiple discharge openings in the nozzle body, while a third type impinges small jet streams of water by multiple small opposed openings on the outside of the nozzle body.

Spray nozzles are available that not only resemble regular pendent sprinklers, but also function as a regular sprinkler-the only difference is the specially designed deflector. These spray nozzles may have a fusible link similar to a standard sprinkler head.

There are many other approved spray nozzles on the market, but one unique nozzle deserves mention. The spiral-type spray nozzle is designed to discharge water along the axis of an external spiral with a diminishing inside diameter. The water discharging over and around this spiral literally peels off in thin layers from the surface of the cone, and this thin layer of water breaks into a spray as it leaves the spiral.

It is usually a requirement of the insurance interests and is always a recommendation of the fire protection engineer, to install a strainer in the main water supply to a water spray system to remove any foreign material such as sand, dirt, or scale before it lodges in the small orifice of a spray nozzle and disrupts the discharge pattern. Water spray nozzles that depend on very small jet stream orifices to create a specific discharge pattern will have their own individual internal strainer at each nozzle, but this does not eliminate the need for the main line strainer to remove larger foreign material before it can clog the nozzle strainer. Main line strainers are designed so they can be flushed periodically.

In addition to the possibility of internal clogging of small orifices, spray nozzles installed in an atmosphere similar to paint vapor or a corrosive atmosphere may require blow off caps. Blow off caps fit over the nozzle in such a manner that the water pressure will immediately blow them off. To prevent personnel from being struck by the blown off cap, it is attached by a chain or cable to the nozzle body. This attachment allows for reuse after the system is shut off.

Configuration of Equipment

Once a nozzle is selected, the next step is to study the physical arrangement and configuration of the structure or equipment to be protected. This criteria will enable the engineer to determine the proper placement of the nozzles and the angle at which the nozzles will be installed, since each nozzle has a predetermined discharge cone or pattern. The nozzle discharge capacity, location, and spacing of the nozzles, angle of discharge, and the nozzle discharge cone or pattern, all contribute to the delivery of the desired density requirement over the entire surface to be protected.

Water Supply

Hydraulic calculations based on supply and demand will determine the proper size and water requirements. Water supply needs must take into consideration the number of nozzles that will operate in any anticipated fire condition, and the required duration of the water supply to the water spray system, and this must be evaluated for each installation. The anticipated duration of discharge depends on the nature of the hazard and the purpose for which the system is designed. In discussing the water supply, two factors warrant discussion. First, the number of nozzles that will operate. Most water spray systems are open head deluge type, but closed orifice nozzles are available. These systems operate like sprinkler systems, with only the involved nozzles operating. Secondly, the size of the system. Insurance standards recommend that the size of water spray systems be limited so that the designed discharge rate, calculated at the minimum pressures required for the nozzle discharge to be effective, will not exceed 3000 gpm.

Sizing the System

Naturally, the size of a water spray system is governed by:

- The nature of the hazard
- The combustibles involved
- Size of the area being protected
- Type and amount of equipment or structural elements being protected
- The size of the potential fire hazard

Other systems in the area could limit the size or spread of a fire. Fire wall subdivision of the area would afford containment of a fire, and dikes, curbs, and drainage facilities can limit the spread and extent of a flammable liquid fire. Flow from a water spray system can vary from a few hundred to several thousand gallons per minute depending on the type of hazard and the extent of the anticipated fire area. The duration of the flow from a water spray system depends on many factors and can vary from 2 min to more than 2 h when exposure protection is involved. Water spray systems usually protect high value and high hazard equipment or facilities, and the detection system that activates these systems must be reliable, dependable, and in some cases, quite sensitive.

Water Mist Systems

Water Mist has been demonstrated to be a suitable system replacement for gaseous suppression systems (Dry-Agent Automatic Suppression Systems) in many commercial and industrial applications. Like water spray it is:

- Environmentally acceptable
- Non-toxic
- Safe for use in areas occupied by humans
- Highly efficient as a fire suppression agent
- Low electrical conductivity

The design of water mist systems in the US is governed by NFPA 750. The droplet size and therefore the cooling influence of the water emitted from such systems is superior even to water spray. Disadvantages of the systems include cost due to the special pumps involved and the extensive use of expensive stainless-steel components and piping. Another disadvantage may be the large power draw of the high-pressure pumps.

However, when large areas (such as ships, and underground structures) need to be protected with a minimal quantity of water these systems have proven to be a superior alternative.

Extinguishing Method

Water mist also cools the fire to such an extent that it collapses the heat portion of the fire triangle. Larger drops have larger surface areas than small drops. This is important because it is through the surface area that the water absorbs heat. The more surface area that can be exposed to a fire, the better and quicker the extinguishment will be. In one drop of water, we can normalize the surface area to 1 and vaporization occurs in about 1 s. Class 2/3 mist has 40 drops in the same volume, and this results in a surface area be exposed to the fire of 10 time that of a sprinkler droplet, and a vaporization rate of 0.1 s.

Electrical Conductivity

Tests have been run at very high voltages from nozzles onto plates with demonstrating that water mist is not a good conductor of electricity.

Design Considerations

Water mist systems generate particularly small water particles (on the order of 0.3–0.0005 mm). These particles require high operating pressure pumps and therefore large power supplies to produce these small water particles. In addition, due to this high pressure, it is necessary to use high pressure stainless steel piping, which is very expensive, but easily installed due to its flexibility.

As it will not splatter fires or flammable liquids, water mist be used on all four classes of fire, although design considerations, such as initial fire size may be a limitation.

Water removes oxygen (and steam) at the precise point of combustion, so it is not harmful to humans, and does not require the space where it is deployed to be evacuated.

Water Supply

Use of a particular local water supply can depend on the particular suspended particle measurement of the supply. Areas with a large quantity of suspended particles in their water supply can clog screens required to be located at the entrance of the heads.

References

Hall, John R. "U.S. Experience with Sprinklers and Other Automatic Fire Extinguishing Equipment." (2010)

Golinveaux, James. "Quick Opening Devices (Dry Pipe Valve Accelerator)." 3/12 2018. <http://www.tyco-fire.com/TFP_common/accelerators-white.pdf>.

Hydraulic Calculations of Sprinkler Systems

10

Hydraulic Calculations

Automatic sprinkler systems are classified as light hazard, ordinary hazard, and extra hazard based on the building use or type of occupancy the system it is designed to protect, and there are different pipe size schedules for each of these three classifications.

In NFPA 13 (2016) the design criteria used in hydraulic calculations is also based on three classifications, but ordinary hazard classification is subdivided into Ordinary Hazard Group 1 and Ordinary Hazard Group 2. Extra hazard is also divided into Extra Hazard Group 1 and Extra Hazard Group 2 (See Fig. 10.1).

When using the pipe schedule to size sprinkler piping, a table in NFPA Standard 13 provides the requirements for water supply. This table is broken down into requirements for water supply, and requirements for light and ordinary hazard, Groups 1 and 2. Extra hazard requirements for water supply must be approved by the insurance carrier.

When using hydraulic calculations to size the system piping, another table in NFPA Standard 13 can be used to obtain the design criteria for the different classifications, or the design criteria will be provided by the insurance carrier. Since the piping schedule method is rarely used, this chapter addresses the hydraulic calculation method for sizing pipe. The purpose of this chapter is not to instruct the reader in how to make hydraulic calculations, but to present the concepts so as to provide familiarity with the procedure and take out some of the "mystery."

Design Criteria

By hydraulically calculating the area of operation, one is saying that the pipe is sized to deliver the required flow and pressure each head in the area of operation and sizing the pipe back to the system water supply in order to deliver the demand for the area of operation.

Hydraulic calculations use design criteria which is based on the type of occupancy, and mandates that each head on the system will discharge not less than a specified, or predetermined, number of gallons per minute per square foot of the protected area. For closed head sprinkler systems, there must also be an area of operation, or area of application, which is the maximum square footage a fire in that type of occupancy has the potential to cover. This area establishes the quantity of sprinklers required open to contain and extinguish the fire within the area of operation. This area also establishes the number of heads to be hydraulically calculated:

> Divide the square footage of the area of operation by the square footage each sprinkler head covers. If each sprinkler head covers 100 sq. ft, and the area of operation is 3000 sq. ft, 30 heads will have to be calculated.

Over this area of operation, every sprinkler head must have sufficient volume and pressure to discharge at least the predetermined and stipulated gallons per minute per square foot. This area of operation must be the area that is furthermost removed, hydraulically, from the source of supply.

© Springer International Publishing AG, part of Springer Nature 2019
R. C. Till, J. W. Coon, *Fire Protection*, https://doi.org/10.1007/978-3-319-90844-1_10

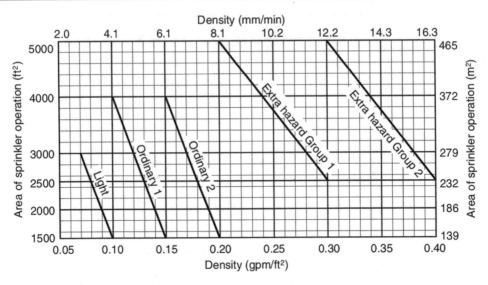

Fig. 10.1 NFPA 13 Density/Area curves. (Reprinted with permission from NFPA 13–2016, Standard for the Installation of Sprinkler Systems, Copyright © 2015, National Fire Protection Association, Quincy, MA. This reprinted material is not the complete and official position of the NFPA on the referenced subject, which is represented only by the standard in its entirety which may be obtained through the NFPA website at www.nfpa.org)

This is commonly referred to as the most hydraulically remote square feet. If the gpm per square foot can be accomplished by hydraulic calculations in the most remote area hydraulically from the source of system water supply, all the heads in the other areas of operation (closer to the system water supply) will discharge the required gpm per square foot, or more, since the pipe friction loss to reach these closer areas of operation is less than that required to reach the remote area of operation. For example (Fig. 10.2):

> The design criteria for this facility is 0.30 gpm per square foot over the most hydraulically remote 3000 sq. ft. Therefore, it is only necessary to determine which 3000 sq. ft are the most remote, hydraulically, from the system water supply. With each head covering 100 sq. ft, hydraulically calculate the 30 heads in this area. As mentioned previously, all the other areas of operation (the 3000 sq. ft) closer to the supply will receive the same, if not more, than the water demand in the remote area.

As stated, hydraulically calculating the heads in the remote area of operation means sizing the piping to these heads so that each head will be discharging not less than the design criteria mandates, namely 0.30 gpm per square foot or in the example with 100 sq. ft per head, 30 gpm per head (0.30 times the 100 sq. ft).

When calculating a deluge system, the design criteria will state so many gpm per square foot over the entire area, since every head on a deluge system is the entire area covered by the system.

Flow Data

In order to hydraulically calculate a system, the flow data must be available. Flow data is the supply criteria. This is the static pressure, or the pressure in the supply main with no water flowing. In addition, the residual pressure is that left in the water supply main when water is flowing. But residual pressure is of no value unless the amount (gpm) of water flowing at the time residual pressure is obtained is known. It is imperative to obtain all three facts:

- static pressure
- residual pressure
- flow in gallons per minute when the residual pressure was read on the gauge.

All three of these items are necessary to prepare a supply graph. Figure 11.3 indicates the supply graph for a water supply from a source like a city

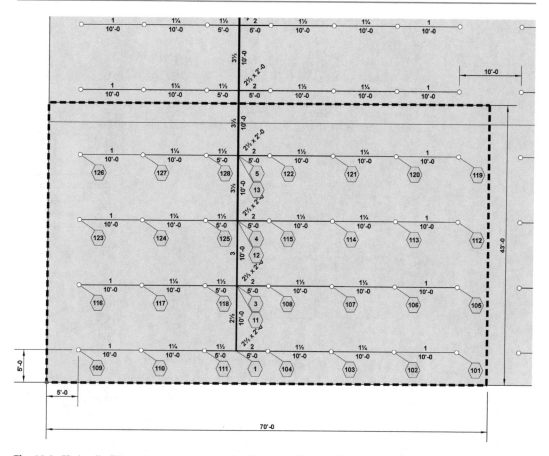

Fig. 10.2 Hydraulically most remote area – Curfew Company diagram. (Courtesy of MEPCAD)

water main-it is a straight line. When the system demand is calculated, it is plotted on this graph. If the demand is below the supply line, the supply will satisfy the demand, but if the demand is above the supply line, the supply will not satisfy the demand. It is always recommended and should always be specified, and is sometimes required by the authority having jurisdiction, to have the calculated system demand fall at least ten psi below the supply line. This is important because flow data on a water main can depend on peak usage hours. For this reason, the information sheet on a set of hydraulic calculations that describes the flow data must include date, location, day of the week, and even time of day.

Flow data obtained from a flow test on a city main performed at ten o'clock on a Sunday would obviously produce much better results than if the same flow data were obtained at 10 o'clock on a Monday when domestic and industrial use of

water was at a much greater level. This example is an exaggeration, but it illustrates that flow data can change during the course of a day or week, and the 10 psi safety factor of the demand below the supply curve allows for these periodic "normal" changes in the flow data. Keep in mind that the supply curve for a fire pump actually includes the pump curve, or a combination of the two, which reflects what residual pressure the pump suction is receiving when it is delivering the required demand volume.

Flow Test

Flow data from a municipal or private water supply is obtained by conducting a flow test. A flow test consists of placing a pressure gauge on one 2–1/2″ outlet of a hydrant and opening the hydrant nut to admit water. No water is flowing because the

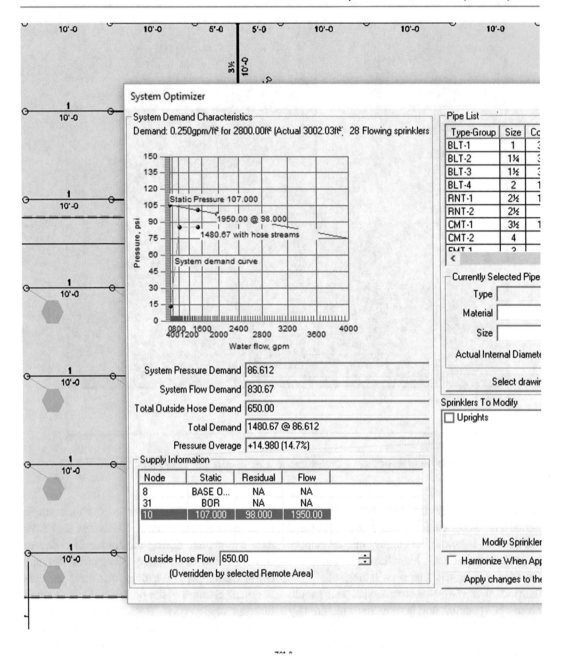

Fig. 10.3 Pressure and Flow Demand Using Autosprink Software. (Courtesy of MEPCAD)

pressure gauge is screwed into the hydrant outlet cap, and the other caps on the hydrant are still in place. With no water flowing from this hydrant (it is now under pressure), the static pressure is indicated on the gauge.

One person is stationed at this hydrant with the pressure gauge, and another goes to a second hydrant on the water main, and, with the hydrant still closed, removes a 2–1/2″ hydrant cap. When both individuals are "ready," the second person opens the hydrant to allow water to flow from the open 2–1/2″ hydrant outlet. When this individual notices that the flow from the hydrant is constant, a pitot tube is inserted in the flow. This tube measures the velocity pressure of the flowing water, which is recorded on a pressure gauge attached to the pitot tube. When the person feels that the reading on the tube pressure gauge has stabilized, a signal is given to the other person at the static pressure gauge hydrant, and the static pressure gauge, which is now registering residual pressure, is recorded.

When the test is over, and the hydrants returned to their normal condition, the reading on the tube pressure gauge is compared with a chart which converts velocity pressure to flow in gallons per minute. The flow data is complete: static pressure, residual pressure, and flow in gpm at the residual pressure.

This is a simplified explanation of a flow test. There are certain items that must be taken into consideration, such as the size and shape of the hydrant outlet. These factors influence the velocity pressure of the hydrant discharge, and the chart that converts velocity pressure to flow has allowances for these conditions in the form of different columns on the chart, or a factor to multiply the velocity pressure by to obtain a true reading.

Approximate Calculated Demand

The approximate gpm demand of a system can be obtained from the design criteria by multiplying the density, gpm per square foot, by the area of operation. To bring this figure up to a more accurate demand figure and closer to what will be the gpm at the bottom of the hydraulic calculation sheet, multiply the answer by 1.15. The friction loss in the pipe in the calculations will increase the psi at each head and subsequently result in slightly more gpm at each head. The end head in the area of operation where the calculations begin is the only head that will discharge exactly (in our example) 30 gpm–0.30 times the area covered by head, namely 100 sq. ft. In order to discharge 30 gpm, this head requires a specific amount of pressure, in this case approximately 29 psi. In getting 30 gpm through the 1″ pipe 10′ long there will be almost a three psi loss due to friction in this length of pipe. Therefore, at the second head, the psi will have to be 29 psi plus three psi, or 32 psi to deliver 30 gpm to the end head at 29 psi. This means that the second head will be discharging at 32 psi, approximately 32

gpm. This increase in pressure at each succeeding head increases its discharge. The example above is extreme, since the friction loss of approximately three psi is in all probability too great, and the pipe size would have been 1–1/4′ to reduce the friction loss to 0.74 psi in lieu of 2.76 psi with the 1″ pipe. This does point out the necessity to adjust pipe sizes in performing hydraulic calculations in order to maintain the demand, both in volume and pressure, remain below the supply curve. The foregoing illustrates the reason to increase the design criteria by 15% and even 20% to obtain a realistic demand figure. As the stamp on a preliminary plan states, this figure is preliminary and not for construction. The only figure that can be used to size fire pumps, storage volumes underground mains, etc., is the figure at the bottom of the hydraulic calculation sheets.

Computer Versus Hand Calculations

Most hydraulic calculations for fire suppression systems are currently performed by computers. The extremely simplified example presented later in this chapter points out one minor reason for performing hydraulic calculations by use of a computer program. Had this calculation been performed by hand, and the excessive friction loss been reflected in too small a pipe sizing throughout the calculation, the demand would probably have been above the supply curve, the calculation sheets thrown away, and the calculations redone with adjusted pipe sizing to reduce friction loss. By computer, these adjustments could be made, and the calculations redone almost at the flick of a few keys.

The hydraulic calculations prepared by hand that have been included in this text are merely presented to show the steps involved in preparing hydraulic calculations. Including these calculation sheets is basically an effort to establish the procedure for calculating a sprinkler system by "walking the reader through" a sprinkler calculation to offer the basic concept of hydraulic calculations. The hand calculations are based on a common sprinkler layout: One common supply main (the trunk) feeds the sprinkler lines (the branches).

Sprinkler systems where the lines are all joined together are known as *gridded systems*. These systems allow much smaller pipe sizes but are next to impossible to hydraulically calculate by hand. The calculations must be done by computer (unless performed by a highly experienced engineer with a lot of time on his hands).

Necessity of Hydraulic Calculations

Unless the system is laid out, hydraulically calculated, and the pipe sizes derived from the calculations indicated on the plans, a detailed cost estimate cannot be prepared. If the design criteria is given but a preliminary system design with pipe sizes is not presented, the estimator will not have any concept of pipe sizing. Many performance specifications will provide the design criteria, but because it is a performance spec, no preliminary system design will be offered, and the estimator is at a loss for pipe sizes and lengths. Caution: If the engineer on a project elects to use his or her experience and judgment in selecting the design criteria for the hydraulic calculations from the NFPA Standard 13 table, it is always advisable to verify the design criteria with the insurance carrier. Insurance carriers know the operation and hazards of their clients, and if the engineer selects a design criteria that is not adequate to satisfy the insurance requirements, this can cause additional and unnecessary expense in recalculating, redesign, and possibly retrofitting the installation.

Summary

Hydraulically calculated systems provide assurance that an efficient and effective amount of water will be discharged on the combustion. With a pipe selection system, it can only be assumed that the combustion will be suppressed and/or extinguished by the system discharge. There is an industry adage that if you give five sprinkler designers a building to be protected, you will get five different system designs. The same holds true of hydraulic calculations and the resultant pipe sizing. The previously mentioned performance specification is a variation in layouts and calculations. The performance system and specification leaves the design up to the bidder or contractor. Usually only the point where the system water supply enters the building is indicated, and the design, system criteria, and parameters are covered in the specifications. This can sometimes lead to a wide variance in bids because one bidder may have figured a gridded system and another a "Christmas tree" system. In addition, the architect and engineer generally have little control over the system layout. When shop drawings (installation plans) are received to be checked, the architect/engineer has little recourse as long as the system design meets specifications and satisfies design criteria. The architect/engineer may recommend design alterations if the proposed layout does not meet expectations relative to appearance or potential maintenance or space allocation problems. A preliminary system design with pipe sizes determined by hydraulic calculations will correct some of these problems, and all the bids will be based on the same factors, but the preliminary design and pipe sizes may not be the most economical or efficient. It is, therefore, always recommended that the specifications allow the successful bidder to redesign and recalculate the system based on the specification requirements, subject to the approval of the engineer.

The next portion of this chapter contains Fig. 10.4, Hydraulic Calculation Sheets, and an explanation (Fig. 10.5) of how these figures were derived.

Client __Curfew Company__   Page __1__ of __4__
Project No. __13903.300__   Date __12/90__   Made By __W.C.__
       Checked By __A.N.__
       Preliminary ___ Final __✓__

Hydraulic Design Information Sheet

Name __Curfew Company__   Date __12/12/90__

Location __Kansas City, Missouri__

Building __Industrial - Machine Shop__   System No. __1__

Contractor __B & V Sprinkler Co.__   Project No. __13903.300__

Calculated By __Walter Coon__   Drawing No. __1 of 1__

Construction: ☐ Combustible _____ ☒ Non-Combustible _____ Ceiling Height __25__ Ft.

Occupancy __Machine shop with hydraulic presses__

System Design

☒ NFPA 13:   ☐ Lt. Haz.   Ord. Haz. Gp.   ☐ 1   ☐ 2   ☒ 3   ☐ Ex. Haz.
☐ NFPA 231   ☐ NFPA 231C: Figure _____ ;   Curve _____
☐ Other (Specify) _____
☐ Specific Ruling _____ Made By _____ Date _____

Area of Sprinkler Operation __2800 S.F.__
Density __0.25 GPM/SF__
Area Per Sprinkler __100 S.F.__
Hose Allowance GPM: Inside ⎱ __650 GPM__
Hose Allowance GPM: Outside ⎰
Rack Sprinkler Allowance __Not Applicable__

System Type
☒ Wet   ☐ Dry   ☐ Deluge   ☐ Pre-Action
Sprinkler or Nozzle
Make __Viking__   Model __S.S.U.__
Size __17/32 - inch__   K-Factor __8.2__
Temperature Rating __165° F__

Calculation Summary
GPM Required __848__   PSI Required __86__   At Base of Riser
"C" Factor Used: Overhead __120__   Underground __140__

Water Supply

Water Flow Test
Date & Time __10/90__ __10 a.m.__
Static PSI __107__
Residual PSI __98__
GPM Flowing __1950__
Elevation __0- Ft.__

Pump Data
Rated Capacity _____
At PSI __N.A.__
Elevation _____

Tank or Reservoir
Capacity _____
Elevation __N.A.__

Well
Proof Flow _____ GPM

Location __Hydrant north of new building__
Source of Information __Factory Mutual flow test__

Commodity Storage

Commodity _____   Class _____   Location _____
Storage Height __N.A.__   Area _____   Aisle Width _____
Storage Method: Solid Piled _____ %   Palletized _____ %   Rack _____ %

Rack
☐ Single Row   ☐ Conventional Pallet   ☐ Automatic Storage   ☐ Encapsulated
☐ Double Row   ☐ Slave Pallet __N.A.__   ☐ Solid Shelving   ☐ Non-
☐ Multiple Row   ☐ Open   Encapsulated

Flue Spacing in Inches
Longitudinal _____ Transverse _____   Clearance From Top of Storage to Ceiling
__N.A.__ _____ Ft. _____ In.

Horizontal Barriers Provided _____

A.L.V.	—Alarm Valve	G.P.M.	—Gallons Per Minute	P.S.I.	—Pounds Per Square In.
C	—Cross	GV	—Gate Valve	PT	—Total Pressure in P.S.I.
CV	—Swing Check Valve	LT.E.	—Long Turn Elbow	Q	—Flow Increment
DEL.V.	—Deluge Valve	PE	—Pressure Loss Due	Q	—Summation of Flow
DPV	—Dry-Pipe Valve		to Elevation	ST	—Strainer
E	—90° Elbow	PF	—Pressure Loss Due	T	—Tee
EE	—45° Elbow		to Friction	T.W.	—Thin Wall Pipe

051678

Form TS-FC-2
(Page 1 of 3)

Hydraulic Calculation Sheet # 1

Fig. 10.4 Hydraulic Calculation

Client __Curfew Company__   Page ___2___ of ___4___

Project No. __13903.300__ ___Date __12/90___   Made By ___W.C.___

_____   Checked By ___A.N.___

_____   Preliminary _____ Final ___✓___

17/32" orifice, K= 8.2, .25 GPM/S.F. over 2800 S.F.   100 S.F./Head

Nozzle Type & Location	Flow In G.P.M.	Pipe Size	Fitting & Devices	Pipe Equiv. Length	Friction Loss P.S.I./Ft.	Required P.S.I.	Hyd. Ref. Pt. ⬡	Elev.	Notes
Line # 1	q	1"		LGTH. 10.00	.199	PT 9.30			Q= K√P
①				FTG.		PF 1.99			√P= Q/K
	Q 25.00			TOT. 10.00		PE			√P= 25.00/8.2
②	q 27.55	1¼"		LGTH. 10.00	.210	PT 11.29			P= 9.30
				FTG.		PF 2.10			
	Q 52.55			TOT. 10.00		PE			Q= K√P
③	q 30.01	1½"	T	LGTH. 5.0	.324	PT 13.39			Q= 8.2 √11.29
				FTG. 8.0		PF 4.21			Q= 27.55
	Q 82.56			TOT. 13.0		PE			Q=K√P
	q			LGTH.		PT 17.60			Q= 8.2√13.3
				FTG.		PF			Q= 30.01
	Q			TOT.		PE			
	q			LGTH.		PT			Q= K√P
				FTG.		PF			K= Q/√P
	Q			TOT.		PE			
	q			LGTH.		PT			K= 82.56/√17.60
				FTG.		PF			
	Q			TOT.		PE			K= 19.66
	q			LGTH.		PT			"K" for
				FTG.		PF			Line # 1
	Q			TOT.		PE			19.66
	q			LGTH.		PT			
				FTG.		PF			
	Q			TOT.		PE			
	q			LGTH.		PT			
				FTG.		PF			
	Q			TOT.		PE			
	q			LGTH.		PT			
				FTG.		PF			
	Q			TOT.		PE			
	q			LGTH.		PT			
				FTG.		PF			
	Q			TOT.		PE			
	q			LGTH.		PT			
				FTG.		PF			
	Q			TOT.		PE			
	q			LGTH.		PT			
				FTG.		PF			
	Q			TOT.		PE			
	q			LGTH.		PT			
				FTG.		PF			
	Q			TOT.		PE			
	q			LGTH.		PT			
				FTG.		PF			
	Q			TOT.		PE			

051678

Form TS-FC-2
(Page 2 of 3)

Hydraulic Calculation Sheet # 2

Fig. 10.4   (continued)

Client: Curfew Company Page 3 of 4

Project No. 13903.300 Date 12/90 Made By W.C.

Checked By A.N.

Preliminary _____ Final ✓

Nozzle Type & Location	Flow In G.P.M.	Pipe Size	Fitting & Devices	Pipe Equiv. Length	Friction Loss P.S.I./Ft.	Required P.S.I.	Hyd. Ref. Pt. ⬡	Elev.	Notes
Line #2	q	1"		LGTH. 10.00	.199	PT 9.30			Q= K√P
④				FTG.		PF 1.99			VP= 9/K
	Q 25.00			TOT. 10.00		PE			√P= 25/8.2
⑤	q 27.55	1¼"		LGTH. 10.00	.210	PT 11.29			P= 9.30
				FTG.		PF 2.10			
	Q 52.55			TOT. 10.00		PE			Q= K√P
⑥	q 30.01	1½"		LGTH. 10.00	.324	PT 13.39			Q=8.2√11.29
				FTG.		PF 3.24			Q= 27.55
	Q 82.56			TOT. 10.00		PE			Q= K√P
⑦	q 33.46	2"	T	LGTH. 5.00	.125	PT 16.63			Q=8.2√13.3
				FTG. 10.00		PF 1.88			Q=30.01
	Q 116.02			TOT. 15.00		PE			Q= K√P
Line #1	q 84.54	2½"	T	LGTH. 2.00	.147	PT 18.51			Q= 8.2√16.63
R.P.				FTG. 12.00		PF 2.06			Q= 33.46
	Q 200.56			TOT. 14.00		PE .87			Q= K√P
	q	2½"		LGTH. 10.00	.147	PT 21.44			Q= 19.66 √18.
CM "A"	Q 200.56			FTG.		PF 1.47			Q= 84.54
				TOT. 10.00		PE			
Lines 3+4	q 207.50	3"		LGTH. 10.00	.189	PT 22.91			"Q" For Line
CM "B"				FTG.		PF 1.89			#1 = 84.54
	Q 408.06			TOT. 10.00		PE			PE=(2.0)(.434)
Lines 5+6	q 215.73	3½"		LGTH. 10.00	.203	PT 24.80			Q= K√P
CM "C"				FTG.		PF 2.03			K= Q/√P
	Q 623.79			TOT. 10.00		PE			
Lines 7+8	q 224.40	3½"		LGTH. 6.50	.359	PT 26.83			K= 200.56 / √21.44
CM "D"				FTG.		PF 2.33			
	Q 848.19			TOT. 6.50		PE			K= 43.32
	q	3½"		LGTH. 33.50	.359	PT 29.16			"K" Lines 1+2
BM "E"				FTG.		PF 12.03			Q= K√P
	Q 848.19			TOT. 33.50		PE			Q=43.32√22.9
	q	4"	2 E	LGTH. 111.00	.194	PT 41.19			Q= 207.50
BM "F"				FTG. 20.00		PF 25.41			Q= K√P
	Q 848.19			TOT. 131.00		PE			Q=43.32√24.8
	q	4"	AL.VA.+OS+Y	LGTH. 24.50	.194	PT 66.60			Q= 215.73
Riser				FTG. 22.00		PF 9.02			Q= K√P
	Q 848.19			TOT. 46.50		PE 10.63			Q=43.32√26.8
Inside Hose	q 150.00			LGTH.		PT 86.25	Base Riser		Q= 224.40
				FTG.		PF			PE =(24.5)(.434)
	Q 998.19			TOT.		PE			
Undg.	q	6"	E+4×H FL6	LGTH. 6.00	.036	PT 86.25			PE (6.0)(.434)
F.S. Pc.				FTG. 19.00		PF .90			
	Q 998.19			TOT. 25.00		PE 2.60			
Undg. to cont. City main	q	6"	T+Gate VA	LGTH. 100.00	.036	PT 89.75			
				FTG. 36.00		PF 4.99			
	Q 998.19			TOT. 136.00		PE			
Outside Hose	q 500.00			LGTH.		PT 94.74			
				FTG.		PF			
	Q 1498.19			TOT.		PE			
	q			LGTH.		PT			
	Q			FTG.		PF			
				TOT.		PE			

1500 GPM @ 95PSI

051678

Form TS-FC-2
(Page 2 of 3)

Hydraulic Calculation Sheet # 3

Fig. 10.4 (continued)

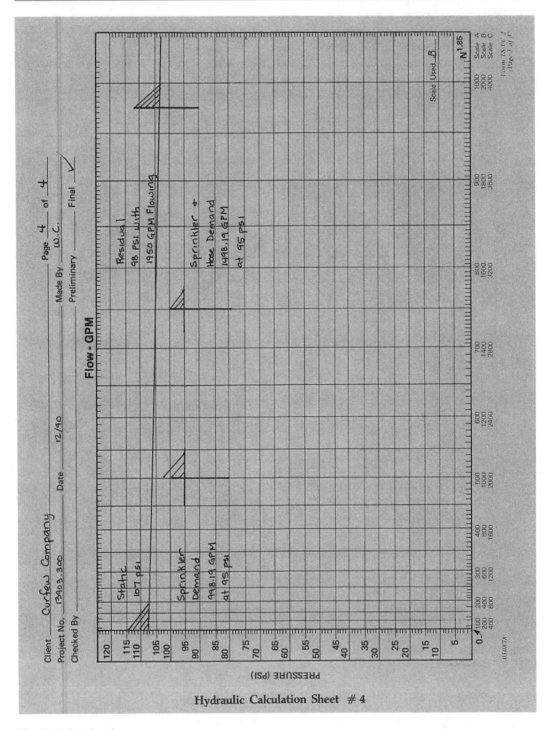

Fig. 10.4 (continued)

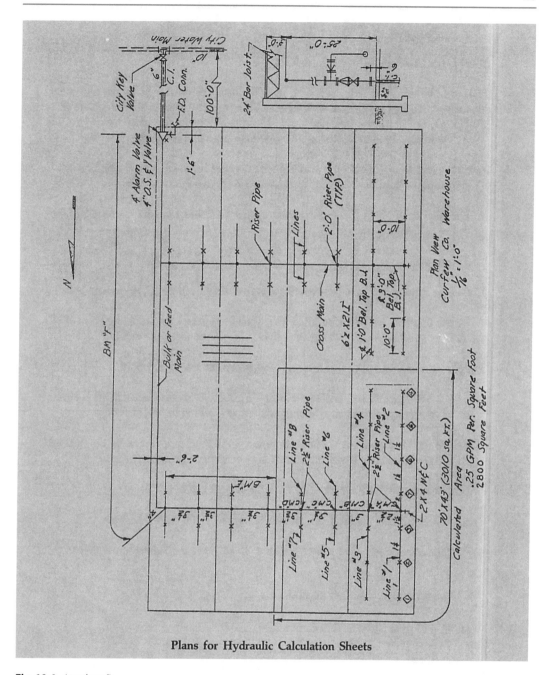

Plans for Hydraulic Calculation Sheets

Fig. 10.4 (continued)

Explanation of Hydraulic Calculation Sheets—1 through 4

A. Sheet One: Cover or Information Sheet.

1. Basic Design Data: This criteria is, in most instances, furnished by the insurance interest having jurisdiction. If density and area are obtained from NFPA pamphlet No. 13, or FM Loss Prevention Data, it is imperative that the criteria be approved by the insurance interest having jurisdiction prior to preparing shop drawings.

2. Flow Test Data: The supply and demand curves (Sheet 4) present a graphic indication of the importance of a realistic and accurate flow test.

3. The balance of Sheet 1 is self-explanatory and furnishes a concise overview of the entire hydraulic situation.

B. Calculations:

1. Design Criteria: .25 gpm per-square-foot area of application—2,800 square feet.

2. Due to construction of the building, the system has been laid out on the basis of 10 feet between heads, 10 feet between lines: 100 square feet per head.

3. An area of 2,800 square feet is selected most remote from the supply.

    a. If the design criteria can be met for this area, there is assurance that areas closer to the supply (shorter supply runs) will be well within the design criteria.

    b. We have picked the "worst condition"—hydraulically—due to the longest supply to feed this area, and is furthest from the system supply.

4. Step No. 1: (2,800 divided by 100 sq. ft./head = 28 heads calculated)

    a. 100 square feet per head .25 gpm per square foot 100 × .25 = 25 gpm.

    b. This means that the furthest sprinkler head from the supply must deliver 25 gpm.

    c. Head 1 (nozzle type and location column).

    d. Q = 25 gpm (flow in gpm column).

5. Step No. 2:

    a. Now we must determine what pressure will be required at head 1 to deliver 25 gpm.

Fig. 10.5 Explanation of Hydraulic Calculation Sheets

(1) In order to keep the pressure in the total system within the flow test range, and keep pipe size to a minimum, and in view of a good gpm flow, we have chosen a "large orifice" head which gives a large volume at a low pressure.

(2) From the manufacturer's data, the L.O. head we are using (Viking) has a "K" factor of 8.2.

(3) The "K" factor is a constant which reduces the friction loss in the head, and fitting the head screws into, to a single number for loss through a single orifice.

b. Using the Formula: $Q = K\sqrt{P}$

Q = gpm flow

K = Constant

P = psi

And substituting

$Q = K\sqrt{P}$

$\sqrt{P} = Q / K$

$\sqrt{P} = 25 / 8.2$

$P = 9.30$ psi

It requires 9.30 psi at the head to deliver 25 gpm. Insert 9.30 in "required psi" column, PT = Total Pressure.

6. Step No. 3:

a. We select 1-inch pipe size for the end head supply and enter this in "Pipe Size" column.

(1) This may have to be changed if the friction loss is too great.

b. There are no fittings to enter in the "Fitting and Device" column since the "K" factor takes into account the fitting supplying the head.

c. We enter 10.00 feet in "pipe equiv. length" column since the pipe supplying head 1 is 10 feet long.

Fig. 10.5 (continued)

    d. This 1-inch pipe must deliver 25 gpm to head 1—the friction loss per foot tables for 1-inch pipe, 25 gpm flowing, is .199 psi/ft. $10 \times .199 = 1.99$ psi (total friction loss)

Enter this in "required psi" column under "PF" (friction loss psi).

    e. The pipe is run level (NFPA 13, 1978, Wet System Piping Drainage) so there is no static, or elevation, loss to enter after "PE" (elevation psi loss or gain).

7. Step No. 4:

    a. Total the psi required (to deliver 9.30 psi at end head) at head 1 and friction loss: $PT = 11.29$

    b. At head 2 there is a total of 11.29 psi and we must determine how many gpm 11.29 psi will deliver.

$$Q = K \sqrt{P}$$

$$Q = 8.2 \sqrt{11.29}$$

$$Q = 27.55$$

    c. Head 2 will deliver 27.55 gpm and head 1 is delivering 25.00 gpm.

    d. Enter head 2 delivery in flow column under small "Q" and add to head 1 flow.

    e. Total "Q" is now 52.55 gpm and the second piece of pipe must deliver 52.55 gpm.

    f. Select 1-1/4-inch pipe. Length 10.00 feet. No fitting.

    g. Friction loss per foot—1-1/4-inch pipe flowing 52.55 gpm is .210 psi/ft. $10 \times .210 = 2.10$ psi friction loss—enter in "PF" and add to obtain a new total "PT" of 13.39 psi.

8. Step No. 5:

    a. Determine what gpm head 3 will deliver at 13.39 psi.

$$Q = K \sqrt{P}$$

$$Q = 8.2 \sqrt{13.39}$$

$$Q = 30.01 \text{ gpm}$$

Fig. 10.5 (continued)

b. Enter this after small "Q" and add to gpm for head 1 and 2.

c. Total gpm that the third piece of pipe must deliver—large "Q"—is 82.56 gpm.

d. Select 1-1/2-inc pipe. Length is 5 feet to the tee on top of the riser pipe. (The riser pipe supplies lines number one and two).

e. The water coming up the riser pipe must change direction in the tee on top to enter line number one.

f. To obtain the friction loss due to this change in direction we look on the "Equivalent Pipe Length Chart." Under "tees" for 1 and 1/2-inch (use the pipe size of the pipe entering the tee) the friction loss in the tee is 8 feet. Enter this 8-feet in "ftg" slot and add to pipe length. Total is 13-feet.

g. From the table 82.56 gpm flowing through 1-1/2-inch pipe—.324 psi friction loss per foot.

$13 \times .324 = 4.21$ psi (Total Friction Loss).

Enter this in "required psi" column under "PF".

h. The total pressure at the centerline of the riser pipe tee is 17.60 psi.

9. Step No. 6:

a. The same process is performed for line number two (Sheet 3).

b. We find that at the centerline of the riser pipe tee the following condition exists.

Line (1) 82.56 gpm @ 17.60 psi.

Line (2) 116.02 gpm @ 18.51 psi.

c. Since it is impossible to have two pressures at this point, and since the minimum pressure to supply line number is 18.51 psi, we must determine what gpm line one will receive at 18.51 psi in lieu of 17.60 psi.

d. To do this we must convert line one to a "K" factor.

$$Q = K\sqrt{P}$$

$$K = Q\sqrt{P}$$

$$K = 82.56\sqrt{17.60}$$

Fig. 10.5 (continued)

K = 19.66 (K for line # 1)

e. Now we can determine what this "single orifice" (line one) will receive (gpm) at the higher psi.

$$Q = K\sqrt{P}$$

$$Q = 19.66\sqrt{18.51}$$

Q = 84.54 gpm (Flow to line 1 at 18.51 psi.)

f. At a balanced pressure (18.51 psi) at the centerline of the tee on top of the riser pipe (2 × 1-1/2 × 2-1/2) the gpm to line (1) is 84.54 gpm and to line (2) is 116.02 gpm.

10. Step No. 7:

a. The riser pipe ("R.P." Sheet 3) must supply line one and line two.

b. Enter line on gpm in "flow in gpm" column (Sheet 3) and add to line two flow. Total "Q" = 200.56 gpm.

c. Select 2-1/2-inch for riser pipe size.

(1) Once again—this is a matter of judgment, taking into consideration total flow vs. friction loss.

d. The length of the riser pipe is 2 feet.

(1) Section shows 3 feet to the cross main and plan indicates centerline of lines 1-foot below top of B.J. Therefore, 2-foot riser pipe.

e. The water from the cross main must change direction to enter the riser pipe, so we have a fitting friction loss to consider. From the table 2-1/2″ tee = 12 feet.

2 feet + 12 feet = 14-foot total, for riser pipe.

f. 200.56 gpm flowing through 2-1/2-inch pipe—.147 psi friction loss/foot 14 × .147 = 2.06 psi friction loss in riser pipe (PF = 2.06).

g. There is a 2-foot elevation change—loss, since the water must rise 2 feet from the cross main to lines one and two.

2 × .434 = .87 psi static loss

Fig. 10.5 (continued)

PE = .87

h. Adding PT at top of riser pipe to PF and PE = a PT of 21.44 psi at the centerline of the cross main tee feeding the riser pipe.

11. Step No. 8:

a. Cross main "A" must supply 200.56 gpm.

b. Length cross main (CM) "A" = 10 feet.

c. Friction loss in selected pipe size for CM "A", 2-1/2-inch, with 200.56 gpm flowing is .147 psi/ft.

$10 \times .147 = .147$ psi.

d. 21.44 psi (PT) plus PF 1.47 psi = 22.91 psi (PT at centerline of cross main tee feeding riser pipe and lines three and four.

12. Step No. 9:

a. Lines three and four supplying riser pipe are all identical to lines one and two and R.P. for one and two.

b. Necessary to determine the gpm lines three and four will receive at a psi of 22.91.

c. Line one and two and R.P. must be reduced to a "single orifice" flow ("K").

$$Q = K \sqrt{P}$$

$$K = Q \sqrt{P}$$

$$K = 200.56 \sqrt{21.44}$$

K = 43.32 ("K" at centerline of cross main tee for R.P. to lines one and two).

d. Substituting

$$Q = K \sqrt{P}$$

$$Q = 43.32 \sqrt{22.91}$$

Q = 207.50 gpm ("Q"—flow—to lines three and four @ 22.91 psi).

Fig. 10.5 (continued)

13. Step No. 10:

  a. CM B must supply lines one and two (200.56 gpm) and lines three and four (207.50 gpm), or a total of 408.06 gpm.

  b. Select 3-inch pipe for CM B.

  c. Length CM B = 10 feet.

  d. Friction loss with 408.06 gpm—3-inch pipe = .189 psi/ft. $10 \times .189 = 1.89$ psi (PF)

  e. Total (PT) at centerline of CM tee for lines five and six is 24.80 psi.

14. Step No. 11:

  a. Lines five and six and R.P. are identical to lines one and two and R.P. which we reduced to a "K" of 43.32.

  b. At 24.80 psi lines five and six will receive:

$$Q = K \sqrt{P}$$

$$Q = 43.32 \sqrt{24.80}$$

$$Q = 215.73 \text{ gpm.}$$

15. Step No. 12:

  a. CM "C" must now supply a total of 408.06 gpm plus 215.73 gpm = "Q" of 623.79 gpm.

  b. CM "C" = 10-foot length.

  c. Select 3-1/2-inch pipe size for CM "C".

  d. 623.79 gpm flowing through 3-1/2-inch pipe—friction loss—.203 psi/ft.

  $10 \times .203 = 2.03$ psi (PF)

  e. Total pressure (PT) at centerline of CM tee supplying line seven and eight is 26.83 psi.

  f. Lines seven and eight and R.P. identical lines one and two and R.P. "K" for lines one and two and R.P.—43.32.

Fig. 10.5 (continued)

$$Q = K\sqrt{P}$$

$$Q = 43.32\sqrt{26.83}$$

Q = 224.40 gpm delivered to lines seven and eight @ 26.83 psi.

g. CM "D" must supply 623.79 gpm plus 225.50 gpm = total "Q" = 848.19 gpm.

h. CM "D" length—to the perimeter of the calculated area is 6.5 feet.

i. Select 3-1/2-inch pipe for CM "D".

j. 848.19 gpm through 3-1/2-inch pipe—.359 psi/ft friction loss.

6.5 × .359 = 2.33 psi (PF).

k. Total psi (PT) at perimeter of calculated area is: 29.16 psi with 848.19 gpm flowing.

16. Step No. 13:

  a. It has been determined that we must supply the "most remote" (to the supply) area with 848.19 gpm @ 29.16 psi.

  b. It can be easily seen that if we accomplish this criteria we will more than satisfy the design criteria (.25 gpm/sq ft) as we approach the supply.

  c. Therefore: we can consider the CM and bulk main as supply pipe or bulk main (BM).

  d. No friction loss for riser pipe tees or tee that feeds the second CM since the water flows "straight through" with no friction loss.

  e. From the perimeter of the calculated area to the last line tee on the CM is 33.5 feet—BM "E".

  f. Select 3-1/2-inch pipe.

  g. 848.19 gpm flowing through 3-1/2-inch pipe (to the calculated area)—.359 psi/ft friction loss.

33.5 × .359 = 12.03 psi "PF".

  h. "PT" at perimeter of calculated area, 29.16 plus 12.03 = 41.19 psi "PT."

Fig. 10.5 (continued)

i. Since BM "F" is long and friction loss would be considerable, select 4-inch pipe.

j. BM "F" = 111 feet. ELL at north end and at riser (supply riser) where water will change direction. From table 4″ ELL = 10 feet pipe equiv. BM "F" = 111.0′ plus 20′ = 131 feet.

k. 848.19 gpm flowing through 4-inch pipe—.194 psi/ft friction loss.

131 × .194 = 25.41 psi "PF."

l. "PT" 41.19 psi plus 25.41 psi "PF" = 66.60 psi at top of supply riser.

17. Step No. 14:

a. Riser is 24.5 feet high from the flange of the underground to the ELL at the top.

b. The Fire Department tee is a straight through flow—no PF loss.

c. The alarm valve and gate valve (OS&Y), 4-inch size, has an equivalent friction loss of 22 feet of 4-inch pipe.

24.5 plus 22 = 46.5 feet.

d. 848.19 gpm through 4-inch pipe—.194 psi/ft friction loss.

e. 46.5 × .194 = 9.02 psi "PF."

f. Riser is 24.5 feet high.

Static Loss = 24.5 × .434 = 10.63 psi "PE."

g. Total psi (PT) at underground flange is 86.25 psi.

18. Step No. 15:

a. Insurance interest having jurisdiction has allowed us to add in the flow of one 1-1/2-inch hose operating (fed by 1-inch pipe from the sprinkler system) at the base of the riser in lieu of "picking it up" as we calculated back.

b. 848.19 gpm ("Q") plus 150 gpm (inside hose "Q") = 998.19 gpm.

19. Step No. 16:

Fig. 10.5 (continued)

a. 4-inch underground supply to fire extinguishing systems is no longer permitted by insurance services, Industrial Risk Insurers (IRI) or FM, so we must run 6-inch to the City main.

b. The "F.S.Pc." (flange and spigot piece) is 6 feet long and has a standard 6×11-inch flange above the floor.

c. To convert to the 4-inch riser we will use a 4×11-inch reducing flange to connect.

d. Equivalent pipe length for the 6-inch ELL on the bottom of the F.S.P.C. plus the 4×11 flange = 19 feet of 6-inch pipe.

e. 6 feet plus 19 feet = 25.0 feet.

f. 998.16 gpm flowing through 6-inch pipe - .036 psi/ft friction loss.

$25 \times .036 = .90$ psi "PF."

g. F.S.PC. is 6 feet high (long).

$PE = 6 \times .434 = 2.60$ psi static loss (PE).

h. PT = 89.75

i. 100 feet 6-inch to City main.

j. 6-inch gate valve (City key valve-tapping valve) plus City main tee ($10 \times 10 \times 6$) equivalent 6-inch pipe length 36 feet.

k. 998.16 gpm flowing through 6-inch pipe—.036 psi/ft friction loss.

$.136 \times .036 = 4.99$ psi PF

l. 89.75 psi "PT" plus 4.99 psi PF = 94.74 psi "PT."

20. Step No. 17:

a. Add 500 gpm for use of a hydrant while the sprinkler system is in use 998.16 + 500 = 1498.19 gpm.

Summary: Demand 1500 gpm @ 95 psi.

Fig. 10.5 (continued)

Foam Systems

11

History: Foam Systems

In August 27, 1859, Edwin L. Drake struck oil 69–1/2′ below the earth's surface near Titusville, Pennsylvania, and created the first producing oil well in the United States, pumping 25 barrels a day. The discovery ushered in not only a new era in technology, but also the beginning of new and unique fire protection problems.

The oil era actually began in 1855, 4 years before Mr. Drake struck it rich in Pennsylvania, when a chemist named Benjamin Stillman made the first fractional distillation of oil. Stillman's process was initially used to obtain oil for lamps, but this led to the discovery of higher boiling fractions which were useful as lubricants. Later, a lower fraction gasoline would find its application in powering the automobile.

Between the Civil War and the turn of the twentieth century, the chemical industry made tremendous strides and mushroomed into a giant among business enterprises. The fire protection industry, accordingly, had to keep up with the growth of new technology. New products, new processes, and new equipment were developed to control new fire hazards, and this was especially true of the flammable liquids that were developed.

America really entered the liquid fuel age at 10:30 A.M. on the 10th of January in 1901 when Spindletop roared like a cannon and spouted oil a 100 over the top of the derrick. Spindletop was located in Texas, and after drilling through 140′ of solid rock to a total depth of 1020′, Spindletop spewed 800,000 barrels of oil in 9 days before it was capped. Spindletop yielded more oil in the first year than the entire world was producing at the time.

More important than the statistics of this first oil well in Texas was the fact that Spindletop revealed to the world the nations greatest oil field, ushered in the modem petroleum industry, and opened the door to the liquid fuel age. Oil discovery in Texas revolutionized industry and raised the nations standard of living.

Flammable liquid fire protection during the years following the success of Spindletop was literally a cannon maintained in a loaded and ready condition. A fire in a holding tank of oil resulted in the firing of the cannon to shoot holes in the tank or blow it apart so the fuel could fill the dike where it could burn itself out.

In 1896, there were four operable automobiles in the United States, and by 1911 there were more than 600,000. In 1903, Orville Wright successfully piloted the Flyer at Kitty Hawk, and the age of aviation was born. The value of petroleum and its usefulness increased as automobiles and airplanes developed and multiplied. The other uses for petroleum rapidly made this resource the starting point for thousands of hydrocarbon-based products.

Naturally, flammable liquid fires were common, particularly when the internal combustion engine became practical, creating a tremendous market for gasoline. Water was applied to these

© Springer International Publishing AG, part of Springer Nature 2019
R. C. Till, J. W. Coon, *Fire Protection*, https://doi.org/10.1007/978-3-319-90844-1_11

flammable liquid fires, but this traditional firefighting agent was not very effective in control or extinguishment. As always, scientists saw a challenge, industry saw a market, and the result was the development of a new flammable liquid fire extinguishing agent – foam. When a flammable liquid bums, it is the vapors from the surface of the liquid that are burning. Foam mitigated this process.

The lowest temperature of a flammable liquid that will produce vapors in enough concentration to form a mixture with air to ignite the vapors at the surface of the liquid is the flash point of the liquid. The flash point of a liquid determines the classification of the liquid as a flammable or combustible liquid. Each of these liquid classifications is subdivided into subclassifications by temperature range and by the boiling point of the liquid, but the basic, conversational, difference is this: A flash point below 100 °F classifies the liquid as a flammable liquid, whereas a flash point at or above 100 °F classifies the liquid as a combustible liquid. This generic classification of flammable and combustible liquids is extremely important when considering an adequate protection agent and system. Consider how easily the fumes from low flash point gasoline can be ignited at room temperature, and how difficult it is to ignite the fumes from fuel oil at room temperature.

Since the vapors given off by the liquid ignite and bum, the best way to extinguish a flammable or combustible liquid fire is to cut off the supply of vapors, to prevent air from mixing with the vapors, or both. Foam is one answer. Foam applied to a burning liquid fire flows over the surface of the liquid and forms a blanket that not only prevents the vapors from leaving the surface of the liquid, but also prohibits the air from reaching the vapors to form an ignitable mixture of vapors and air. When there is a flammable liquid fuel spill which has not ignited, foam applied to the spill will prevent the vapors from leaving the surface and air from mixing with the vapors, should the spill contact an ignition source. This is foam application in the role of fire prevention. Foam is an aqueous (water) suspension of air or gas in the form of small bubbles surrounded by a film of solution. Foam is lighter than the flammable or combustible liquids, so the blanket of foam will float on the surface of the liquid. In addition, there is a certain measure of cooling action by heat absorption from vaporization of the water contained in the very thin films in the foam. This heat absorption is both derived from the fuel surface and adjacent metal surfaces. Some theories consider this cooling action to be a major factor in the extinguishing action of foam, and in some applications, with some types of foam, this may be true. A good foam blanket will strongly resist the heat and flame of the fire, as well as disruption due to wind, thermal drafts, and mechanical agitation, and will reseal an opening in the blanket. One foam (aqueous film forming foam) does not have this foam blanket integrity, but it is far superior to other foams in its ability to accomplish a quick knock-down-extinguishment of a flammable liquid fuel spill fire.

Use and Calculation of Foam

Foam can be used with open head deluge systems, but with careful attention to proportioning (explained in the next section), foam systems can be used with wet pipe, dry pipe, and pre-action systems. With deluge systems every head is discharging, and knowing how much gpm each head is discharging at known pressure, the demand, or total gpm, is easily calculated. Knowing that the demand will be constant, proportioning the foam concentrate into the water supply can be calculated based on this constant demand. The system should be designed for the individual location in which it is to be located.

With a closed head wet pipe, dry pipe, or pre-action system, the demand cannot be accurately determined because the combustion may fuse two heads, 10 heads, or 100 heads. The demand will be based on the number of heads that fuse and discharge foam/water solution. Attempting to accurately proportion the correct percentage of foam concentrate into the water at different flows is an difficult design situation, but proportioning has been made easier over time by electronics.

Some foam deluge systems use spray heads as discharge devices. The discharge pattern from these foam spray nozzles is similar to the sprinkler and foam/water sprinklers that have been described, but these foam spray nozzles are larger and have a heavier discharge.

Proportioning

Basically, foam concentrate is mixed with water to form a foam/water solution which is transported through a piping system to a discharge device or devices where it is mixed with air (aerated) to produce the foam. There are any number of methods of injecting the foam concentrate into the water in a predetermined proportion. The equipment or device to accomplish this action is the proportioning part of a foam system. Foam concentrate is available in a number of concentrations, but the most common are 3% and 6%. When 3% foam concentrate is introduced into the water in the system, the solution that is transported to the discharge device or devices consists of 3% foam concentrate and 97% water. The same proportions are created with 6% foam concentrate, namely 6% foam concentrate and 94% water.

Correct foam proportioning is one of the most important factors in the design of a foam system because the quality of the foam produced depends on the efficiency of the proportioning system. The optimum quantity and quality of foam discharged on a flammable liquid fire must be obtained to produce the desired extinguishing results. Should the percentage of foam concentrate in the water solution be too high, the discharged foam will be too thick and will not flow quickly across the surface of the flammable liquid or around obstructions. In addition, because more foam concentrate is being proportioned into the water than required, the supply of foam concentrate will be depleted too quickly.

Should the foam concentrate be proportioned into the water at less than the required percentage, the discharged foam solution will produce a foam blanket that has less resistance to break down from the heat and flame of the combustion. The water in the foam will drain off faster and cause quicker foam blanket deterioration.

The foam solution is transported through the system piping until it reaches the discharge devices, and these devices perform the function of mixing the foam solution with air to form the characteristic foam blanket. The discharge devices are often regular automatic sprinkler heads. Since the foam solution is aerated, or mixed with air, by being discharged against the deflector of the sprinkler head *without* any special device other than the deflector to aerate the foam solution, these sprinkler heads are referred to as non-air-aspirating foam sprinklers.

There are other foam discharge devices that aerate the foam solution within the shaft of the head. The discharge from these heads against the deflector creates the discharge pattern. These heads are air-aspirating foam discharge heads.

Both the regular sprinkler heads, non-air-aspirating heads, and the air-aspirating sprinkler heads can function as regular sprinkler heads when the foam supply is exhausted. The discharge of water from these heads will produce the typical sprinkler head spray pattern.

Aqueous film forming foam (AFFF) is a much more effective extinguishing agent on flammable liquid combustion when discharged through regular, non-air-aspirating, sprinkler heads. An example is the lower density (gpm/square foot) required when using AFFF discharged through regular orifice non-air-aspirating sprinklers on a deluge system protecting an aircraft hangar against jet fuel spill fires. The density with these sprinkler heads is 0.16 gpm per square foot with AFFF, and 0.20 gpm per square foot with air-aspirating foam/water heads.

Selecting Proportioning Systems and Devices

Efficient foam concentrate proportioning is required to provide quality foam production that insures optimum performance as an extinguishing agent. Proportioning is the heart of a foam system. As with the results obtained from a computer, the results of a foam system are only as good as the material fed into it. A foam system is only as effective as the proportioning system that introduces the percentage of foam concentrate into the system.

The required foam/water solution flow rate to meet the design criteria and the available water pressure are the two most important features in selecting the proper proportioning arrangement. There are certain nationally specified amounts.

Many fire department pumpers are now fitted with automatic foam proportioning and it may not be necessary to use an external proportioning device to achieve a given foam concentration.

Balanced Pressure Proportioning

The balanced pressure proportioning system uses a tank to hold the supply of foam concentrate. The atmospheric foam concentrate storage tank is not a pressure tank, since the foam concentrate does not have to be under pressure with a balanced pressure proportioning system. The tank can be constructed of mild steel, fiberglass reinforced plastic, or plastic. The foam concentrate tank material must be compatible with the type of foam concentrate to be stored in the tank. There are many atmospheric storage tanks on the market but specifying or purchasing one that is not specifically designed and constructed for the storage of foam concentrate will not only lead to expensive system problems, but will reduce the systems suppression effectiveness. The concentrate life will be shortened, but more important, the concentrate could become contaminated and not produce effective foam for fire suppression.

In this system, the foam concentrate is pumped from the storage tank to the proportioning system. The foam concentrate pumps are positive displacement pumps, not centrifugal pumps. The foam concentrate pump size depends on the design rate of flow (gpm) of the concentrate and the type of foam concentrate. With positive displacement pumps, the discharge volume at a specified pressure is constant. In the chapter on pumps, it is obvious that a positive displacement pump discharges on a straight-line pump-constant volume at constant pressure. A centrifugal pump discharge curve indicates an increase in discharge volume as the discharge pressure decreases.

The foam concentrate is pumped through a bypass from the main concentrate discharge line to a device called an automatic pressure balancing valve. This valve, which is also piped into the main water supply line, regulates the foam concentrate pressure to match the water pressure. As the water pressure varies, the pressure balancing valve adjusts the foam concentrate pressure to maintain the required flow of concentrate into the water.

The water supply may be supplying several deluge systems, and as the combustion operates more systems the water pressure will decrease as the volume demand increases. The pressure balancing valve automatically reacts to this pressure change and maintains the flow of foam concentrate consistent with the changing water pressure. Excess foam concentrate above the required design flow is returned to the storage tank.

The foam concentrate discharge line from the pump is connected to a device called a ratio controller, or concentrate metering device, installed in the main water supply line. This device is a carefully designed venturi. (Recalling physics, the water passes through a throat in the venturi. Since the volume of water into the device is the same volume out of the device, the velocity of the water increases as it rushes through the narrow throat. When the velocity increases, the pressure decreases).

The foam concentrate is under the same pressure of the water supply controlled by the pressure balancing valve, and when the water pressure decreases in the narrow venturi throat, foam concentrate is forced into the water stream. With the pressure balancing valve adjusting the foam concentrate pressure to match the water pressure, and the drop in pressure at the throat of the controller, the foam/water solution leaves the outlet of the controller at the correct foam concentrate-to-water percentage, and the foam/water solution is piped to the system discharge devices.

Reserve Systems

On projects where the foam system is protecting high values, or when the insurance carrier requires maximum, uninterrupted system protection, two foam concentrate storage tanks and two foam concentrate pumps may be required. The second reserve tank is permanently piped into the system, and by manual operation of a valve it will become the primary supply when the first tank has exhausted its supply. Should one pump be out of service, the system is still in a functional condition.

Direct Orifice Proportioning

If there are long runs of pipe from the proportioning system to the foam system piping and discharge devices, the transit time of the foam solution may be several seconds. Some flammable liquid fire hazards cannot afford the luxury of waiting several seconds for the foam discharge to suppress and extinguish the combustion. Aircraft hangars are an example.

A jet fuel spill fire beneath a jet aircraft can cause irreparable damage to the aircraft in the hangar if the combustion is not controlled in 30 s and extinguished in 60 s. So allowing several seconds for detection activation, the discharge time of the foam system becomes a few critical few seconds.

Foam deluge systems that protect this type of hazard have a deluge valve located as close as possible to the system that they supply. Water under pressure is supplied to each deluge valve, and foam concentrate is supplied to a small foam concentrate deluge valve installed immediately adjacent to the water deluge valve. The foam concentrate that supplies the foam concentrate deluge valve is maintained at a higher pressure than the water supply to the water deluge valve. The outlet pipe of the foam concentrate deluge valve is piped into the riser of the water deluge valve above the deluge valve. In this foam concentrate line is a carefully calculated orifice plate. The orifice plate has a small hole in the center to allow the flow of foam concentrate. When the foam deluge system detection system is activated, both the water deluge and foam concentrate deluge valves trip simultaneously and foam concentrate is forced through the orifice and into the water flow leaving the water deluge valve. The foam concentrate pressure is higher (by approximately 15 psi)

than the system water pressure in the riser and can therefore be forced into the water stream in the system riser. Based on water and foam concentrate pressure at the two deluge valves, the orifice size is calculated to admit the proper percentage of concentrate into the water.

A device detects the instant drop in pressure on the foam concentrate supply line and water supply line when the deluge valves trip, and this pressure-sensitive device automatically starts the foam concentrate and fire water pump.

In the chapter covering pumps it is noted that some large diesel drive fire (water) pumps require several seconds to come up to speed. The diesel engine must run at full power so the connected fire water pump can discharge at full volume and pressure.

When a hazard requires the discharge of foam without delay, the foam concentrate pump and fire water pump can be started electrically when the first signal is transmitted from the detection system. The pressure-drop pump-starting system is then merely a back-up.

Detection Systems

Foam deluge systems are usually designed to prevent a false dump of the system when the detection system is activated by some non-fire condition or the combustion is large enough to activate a detector, but small enough to be extinguished by hoses or extinguishers. The detection system is arranged so that the first detector to be activated activates all alarms, and the second detector to be activated dumps the deluge systems. To accomplish the starting of the water and foam pumps electrically, the water and foam pump starting sequence would take place when the first detector is activated (The detection systems used in these applications are discussed in the Alarm Devices chapter).

The water fire pump will harmlessly churn, and the foam pump will discharge foam concentrate back to the tank. If the combustion develops and intensifies sufficiently to activate a second detector, water, foam pressure, and volume will be immediately available.

Jockey Pump

In order to maintain the water supply and foam concentrate supply to their respective deluge valves under pressure, both the water line and foam concentrate line will have a small gpm pressure maintenance pump-commonly referred to as a jockey pump. The water jockey pump takes suction on the supply or suction side of the fire water pump and discharges into the supply line on the discharge side of the fire pump check valve. The pressure is maintained in the supply line to the deluge valve by being blocked from passing back through the pump by the check valve. Here, too, the concentrate cannot pass back through the pump because of the check valve. These jockey pumps maintain pressure slightly higher (approximately 10 or 15 psi) than the starting pressure of the fire water and concentrate pump pressure switch.

When the water and foam concentrate deluge valves trip, the pressure drops faster than the jockey pump can compensate for. The pressure drop reaches the pre-set starting pressure of the pressure switch of the fire water and foam concentrate pumps.

Jockey pumps are discussed previously, but water and concentrate jockey pumps should be small volume (gpm) pumps so there is no possibility of making up the pressure loss when the deluge systems trip. This would prevent the fire water and concentrate pumps from starting. The specifications for the foam concentrate jockey pump should specify that the pump parts and construction shall be compatible with the concentrate being used to prevent the concentrate from rapid deterioration of the pump, gaskets, seals, and metal parts.

Diaphragm Pressure Proportioning

The diaphragm pressure proportioning system does not require foam concentrate pumps because it supplies water pressure to discharge foam concentrate into the water supply and produce the foam/water solution.

The concentrate tank (see Fig. 11.1) is a metal pressure vessel containing a collapsible bladder, and thus it derives the name of bladder tank. This foam system is usually referred to as a bladder tank system. The bladder contains the concentrate, and when water pressure is supplied to the tank surrounding the full bladder the water pressure squeezes the bladder and forces the foam into the piping.

Water Supply

The water supply line is equipped with a ratio controller venturi, and on the water supply side of the venturi a small water line is connected to the water supply side of the bladder tank. The pressure on the outside of the bladder is thereby equal to the water supply pressure. The concentrate bladder has a concentrate line from the bladder to the throat of the venturi of the ratio controller. This line is kept closed by an automatic valve. When the water supply valve is tripped by the system detection system, the valve on the concentrate line to the venturi is opened simultaneously.

Supply water pressure is transmitted to the outside of the bladder in the tank, and foam concentrate is pushed through the concentrate line into the venturi. Here it is proportioned into the water, and the foam/water solution is piped to the system discharge devices. The pressure of the concentrate is equal to the pressure of the water and it can be injected into the water supply where the water pressure drops as the velocity increases at the throat of the venturi.

The atmospheric tank on the balanced pressure proportioning system can be refilled with concentrate during operation of the system, but a bladder tank is a pressure vessel and therefore cannot be refilled during system operation. Refilling the bladder tank with concentrate requires considerable time and care to avoid punching a hole in the bladder. Even a pin hole in the nylon reinforced elastomeric bladder will render the bladder tank useless due to loss of pressure necessary to force concentrate out of the bladder.

Some bladder tanks have a diaphragm instead of a bladder that completely fills the tank. These

Fig. 11.1 Foam concentrate tank and PIV's. (Photograph by Robert Till)

tanks are flanged in the center and the internal diaphragm is fastened to this flange. When water pressure is applied to the diaphragm, which is at one end of the tank, the concentrate is forced through a concentrate line at the opposite end of the tank and thus to the ratio-controller venturi.

High Expansion Foam

The foam that has been previously discussed is called low expansion foam. When low expansion foam is discharged, it forms a layer of foam about 2–3 deep after a 10 min discharge from a deluge system. In contrast, the discharge of a high expansion foam system for 10 min can form a layer of foam 30 high. The expansion ratio of low expansion foam is about 8:1 or 10:1. High expansion foam expansion ratio is about 500:1 or 1000: 1.

High Expansion Foam Generators

To understand how high expansion foam is formed and discharged, recall blowing bubbles through the film on a plastic circle. One bubble

of high expansion foam is similarly produced. A foam concentrate is proportioned into the water, and the foam/water solution is sprayed over a metal or nylon mesh. Air passes through the mesh at a high velocity, creating the high expansion foam discharge. High expansion foam consists of millions of tiny individual bubbles. The foam may be as much as 1000 parts air to one part water. In contact with the fire, the foam changes to steam (which expands 1700 times) and the oxygen content of this steam and air mixture becomes less than 7.5%. The combustion is oxygen starved and extinguished, literally suffocated.

High expansion discharge devices are called high expansion foam generators. These foam generators are usually installed on the roof or high on the walls so the foam will flow into the protected area. Unlike low expansion foam discharged from sprinklers or nozzles under pressure, the only pressure on the high expansion foam comes from the air pressure through the liquid foam solution on the mesh or screen. By mounting the foam generators at a high point, the discharge can use gravity to flow into the hazard area (Fig. 11.2).

Fig. 11.2 High Expansion Foam Generation. (Courtesy of Doug Nadeau)

High Expansion Foam Fire Suppression

High expansion foam also assists in controlling and extinguishing combustion by diverting the flow of air into the fire area by the mass of the high expansion foam discharge. The heat from the combustion forms a thermal column that sucks air into the fire area. The high expansion foam discharge, by its volume, disrupts this flow of air. In addition, high expansion foam has a cooling action because of the discharge water content. This water held in suspension cools the surface of everything it comes in contact with in the hazard area. The bubbles in high expansion foam are relatively uniform in size, the walls are quite tough and elastic films, and there is very little liquid trapped in the bubble to weaken and distort its configuration. All these elements produce a foam with very stable liquid retention, important to cooling action and extinguishment.

Blessing and Curse

High expansion foam can be a blessing and a curse to firefighters. It can form a barrier against the radiant heat of a fire as firefighters advance toward the combustion, and firefighters can actually work surrounded in a blanket of high expansion foam because they can still breathe while completely covered by the foam.

The curse is that although people can still breathe in a high expansion foam blanket, vision is limited to almost zero. Firefighters (and occupants) can trip over unseen objects, fall in pits, or bump into objects.

High Expansion Foam Uses

High expansion foam has many uses and is an effective suppression and extinguishing agent on both ordinary combustibles such as paper and wood (Class A combustibles), and flammable liquids (Class B combustibles). This foam, (commonly referred to as High X foam), is most effective when used in enclosed or confined areas, because when used in an open area it will tend to flow out of the hazard area. If there is wind or a severe draft, it will blow away. However, high expansion foam is very effective fire protection against jet fuel spill fires in aircraft hangars, even with large hangar doors open. When used in aircraft hangars, the discharge of foam must be adequate to cover the hangar floor, build up at the rate of three per minute, and discharge for 10 min. High X foam has proved to be efficient protection in roll paper storage warehouses as well. The foam discharge fills all the spaces between the rolls of paper and eventually covers the tops of the rolls. Sprinkler discharge would soak the paper rolls and ruin the paper stock. A more serious consequence would be the swelling of the paper rolls as they become water soaked. This occurred in a sprinklered roll paper warehouse of a newspaper company, and the paper rolls exposed to the heavy sprinkler discharge expanded and pushed out the concrete block walls. The consistency of High X foam does not soak the material with which it comes into contact. The exterior paper skin may be wet, but the roll is not soaked.

Sizing

To represent the size of a high expansion system, an installation in a high grade rolled paper warehouse 100' long, 55' wide, and 50' high consists of:

Four high expansion foam generators. Each generator is capable of producing 13,500 cubic feet per minute of foam, for a total of 54,000 cubic feet per minute of foam.

All generators are mounted on the roof. This is one very important consideration with this installation, and every high expansion foam system discharging into an enclosed area. The air in the enclosure must be vented because the foam occupies so much volume. The described system is an excellent example-54,000 cubic feet per minute discharge of High X foam will displace 54,000 cubic feet per minute volume of air in the enclosure. The described installation consisted of eight good-sized roof vents to relieve the displaced volume of air.

Advantages

One advantage of high expansion foam is that it uses very little water and a conservative quantity of foam concentrate. The system consisting of the four large foam generators has a foam concentrate supply of 400 gallons, which is sufficient to fill the warehouse four times with high expansion foam.

Activation

High expansion foam systems are usually activated automatically by a signal from a detection system. Actuation may also be manual, and most installations have both automatic and manual activation. When the high expansion foam system is activated automatically, the detection system should set off an evacuation alarm, so personnel can at least commence exiting before the High X discharge begins. The detection system should also close doors, and when the discharge is in an enclosed area, a signal from the detection system opens the vents.

Air Supply

The foam generators can be either electrically or hydraulically operated, and it is imperative that the air supply to the generator be clean air, free of contamination from the combustion. If air supply to the generator could possibly be contaminated with fumes that would affect foam production, or products of combustion, it may be necessary to supply clean air to the generator through a duct.

Types of Foam – Properties Performance

The properties and performance of a foam determine its selection for a specific application. There are numerous types of foam, and the following description of the most common foam types will help provide an understanding of their selection for a specific hazard situation.

Protein Foam

Protein foam is composed of naturally-occurring sources of protein-hoof and horn meal, feather meal, etc. The natural protein material is processed, and stabilizers and inhibitors added to resist bacterial decomposition, protect against freezing, and prevent corrosion.

Protein foams are not as easy flowing on a fuel surface as other low expansion foams, but protein foams do have excellent resistance against the foam blanket breaking up from radiant heat and reignition of the fuel.

At one point these foams were considered obsolete, but they have been reconsidered due to the environmental impacts of other foams. In short, protein foams are biodegradable. Considerations about spores such as those that cause "mad cow disease" may still be a concern.

Protein foams are applied on class B fires, and have a low expansion ratio, with mixing ratios of 3% or 6%.

Fluoroprotein Foam (FP)

This type of foam is similar to protein foam but has the addition of fluorochemical surfactants which produce an easier flowing protein foam. The addition of this surfactant not only allows the production of the air-excluding blanket of protein foam, but also produces a vapor-preventing film on the surface of the fuel.

These foams are also applied on class B fires, and have a low expansion ratio, with mixing ratios of 3% or 6%.

Aqueous Film Forming Foam (AFFF)

Aqueous film forming foam (AFFF) is a synthetic foam concentrate. AFFF is a combination of fluorochemical and hydrocarbon surfactants,

and no protein is involved. The AFFF is totally made up of synthetic products. Surfactants are chemicals that can alter the surface properties of water. Fluorochemical surfactants alter the property of water so that a thin film can spread over the surface of a hydrocarbon fuel even though the aqueous (water) film is denser than the fuel.

These foams are applied on class A and B fires, and have a low expansion ratio, with mixing ratios of 1%, 3% or 6%.

Alcohol-Resistant Foam Concentrate (AR-AFFF)

This is a foam concentrate used for fire suppression on water-soluble fuels that would drain off the water in the foam. Chemicals are added to the foam concentrate so that when the foam is applied to a water-soluble liquid, like methyl alcohol, a membrane is formed between the foam and the water-soluble fuel.

These foams are applied on class B fires, and have a low expansion ratio, with mixing ratios of 3% or 6%.

Medium Expansion Foam

These foam concentrates do not produce foam blankets with vapor reducing films; they produce an aggregate of foam bubbles on the fuel surface. They are used on Class A and some Class B fires. They have a mixing ratio of 2%, 3% or 6%. (When generated from other foams).

High Expansion Foam

As discussed, these foam concentrates do not produce foam blankets with vapor reducing films either; they produce an aggregate of foam bubbles on the fuel surface. They can be used on LNG fires. They are used on Class A and some Class B fires. They have mixing ratios of 1.5%, 2.5% and 3%.

An example of a foam application on a floating roof dam of a fuel tank is shown in Fig. 11.3.

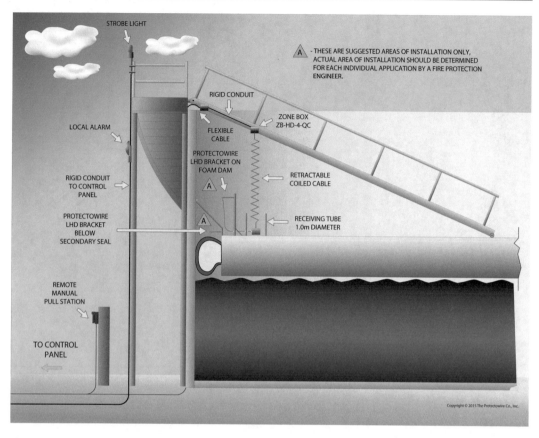

Fig. 11.3 Floating roof tank. (Courtesy of Protectowire)

Dry-Agent Automatic Suppression Systems

12

Dry Agent Suppression Systems

In situations where water and other wet suppression systems are not practical methods of fire protection, dry agent suppression must be used. These include Carbon Dioxide (CO_2), dry chemical, Halon systems, and Halon replacement systems.

There are many instances in which the use of an extinguishing agent can unnecessarily damage valuable equipment that is housed in the protected area. For example, an electrical equipment room, record vault, or storage room requires fire protection because of its operational value, the value of the equipment, and to prevent the spread of fire, yet water discharged on electrical or electronic equipment could considerably damage equipment not involved in the fire. In such cases, many of the agents discussed here can be used to extinguish the combustion without causing unnecessary damage. When applied in the proper concentration, they will put out fires in these applications.

Carbon Dioxide Systems

This section first presents an overview of CO_2 systems, followed by considerations, and then detailed methods of CO_2 system design. Carbon dioxide gas is, in general, colorless and odorless.

Pure CO_2 is a completely odorless, inert, and electrically nonconductive gas which is 50% heavier than air. Carbon dioxide has an history of flammable liquid fire extinguishment because of its ability to form a temporary inert atmosphere above the surface of the liquid. If enough carbon dioxide is introduced into the room to displace enough of the rooms normal atmosphere, the oxygen content of the room can be reduced below the 15% required for support of combustion and the combustion will cease due to oxygen starvation. The normal oxygen content of air is about 21%. Carbon dioxide will not damage delicate, high value equipment or material, nor will it leave a residue.

In addition to its invisible characteristics, carbon dioxide is electrically non-conductive, which makes it an fire protection agent for electrical equipment, particularly for oil cooled and oil insulated electrical equipment (such as oil switches, transformers and circuit breakers).

Carbon dioxide can be used to extinguish fires in all types of combustible materials, with the exception of chemicals and materials containing their own oxygen, which include cellulose nitrate and reactive metals such as sodium, potassium, magnesium and titanium. Carbon dioxide gas is applied in one of two ways: total flooding or local application, depending on the physical characteristics of the room. Total flooding of an enclosed area with carbon dioxide provides three dimensional fire protection. The carbon dioxide gas

© Springer International Publishing AG, part of Springer Nature 2019
R. C. Till, J. W. Coon, *Fire Protection*, https://doi.org/10.1007/978-3-319-90844-1_12

flows in and around all equipment and objects in the enclosed area. This means that equipment can be moved without relocating carbon dioxide nozzles or piping. However, there are instances where it is necessary to protect an isolated hazardous operation, a specific piece of equipment, or hazardous element of a total process, but where total flooding is not practical. Total flooding is difficult in areas that cannot be enclosed to contain the gas. The carbon dioxide system is designed to discharge gas directly onto the hazard to surround it with gas, to render the atmosphere inert and to provide cooling in the immediate area and extinguish the fire.

Local Application Systems

Examples of local application protection include dip tanks, printing presses, motors, and pumps.

The design of local application CO_2 systems is complex, requiring a detailed study, calculations, and design because the area does not remain inert for any appreciable length of time as is possible in a total flooding application. The CO_2 discharged in a local application system is soon dissipated, and its effectiveness therefore decreases rapidly. Detection of the fire and application of the CO_2 must be very rapid so that extinguishment takes place before the surrounding area absorbs sufficient heat to cause re-ignition after the CO_2 has dissipated.

Although in some cases a 30 s burst of CO_2 is suitable, it may be necessary to provide an extended discharge when re-ignition is a serious concern. For example, when there is a flammable liquid and surrounding masses of metal, an extended flow is necessary until the metal is cooled below the ignition temperature of the combustibles.

Other Considerations

The colorless gas quality of CO_2 is often questioned by those viewing a system discharge because it is usually accompanied by a characteristic thick white cloud. Therefore, it is of interest to understand the process by which CO_2 is released into the air. CO_2 is stored and transported through the system piping as a liquid. When this liquid reaches a nozzle, the pressure is released, and the liquid carbon dioxide converts to a gaseous state. The ratio of expansion from liquid to gas is approximately 9:1. The white cloud and a fog-like vapor which may linger for a considerable time consist of solid particles of CO_2 snow, which occurs when the moisture in the atmosphere has been condensed and solidified by the extremely low temperature of approximately $-10\ °F$ experienced when the CO_2 liquid is converted to gas at the nozzle. This low temperature will produce a frost on the CO_2 discharge nozzles, caused by condensation of the moisture in the surrounding atmosphere.

Temperatures

CO_2 discharge temperatures are important to consider in system design. If the CO_2 nozzles are relatively close to metal parts that are in motion, the sudden and severe change in temperature could cause a distortion in the hot metal, causing the disintegration of the machinery and very possibly explosive disintegration in a machine such as a turbine generator. The solid particles of CO_2 evaporate or change from solid to gas without passing through the liquid state very quickly, and in spite of the extremely low discharge temperature, the cooling effect of CO_2 does very little, if anything, in enhancing fire extinguishment. However, it may chill electronic equipment and disrupt its normal operation.

Grounding

Although it is not in a liquid state for long, discharge of liquid carbon dioxide can still produce an electrostatic charge, causing a spark. Therefore, when the CO_2 system is protecting an area that may contain a flammable liquid vapor (creating an explosive atmosphere), metal discharge nozzles should be used, and the nozzles grounded. To dissipate any electrostatic charge resulting from the CO_2 discharge, all objects exposed to the discharge should be grounded.

Total Flooding Systems

The system using carbon dioxide to protect the electrical or computer room previously described is an example of total flooding. The entire room is filled, or total flooded, with a sufficient quantity of CO_2 gas to create an atmosphere which does not contain sufficient oxygen to support combustion. Total flooding of an enclosed area with carbon dioxide provides three-dimensional fire protection. The carbon dioxide flows in and around all equipment and objects in the enclosed area. This means that equipment can be moved without relocating CO_2 nozzles or piping.

There are advantages and disadvantages to using a total flooding carbon dioxide system, all of which are discussed in the paragraphs that follow. However, it must be noted that each hazard situation is unique, and a system should be chosen based on careful study of the location requirements, combined with a careful study of each hazard situation, a thorough knowledge of available extinguishing systems, and a system selection decision that is based on the sum total of all conditions, and not on the hazard only.

Hazards of Total Flooding Carbon Dioxide

Total flooding carbon dioxide systems are restricted from use in normally occupied areas by NFPA 12.

According to NFPA 12 (2015), carbon dioxide acts as both a stimulant and depressant on the central nervous system. Health effects may be seen following exposure to high concentrations of carbon dioxide. Exposure of humans to carbon dioxide concentrations ranging from 17% to 30% quickly (within 1 min) leads to loss of controlled and purposeful activity, unconsciousness, coma, convulsions, and death. Exposure to concentrations from greater than 10–15% carbon dioxide leads to dizziness, drowsiness, severe muscle

twitching, and unconsciousness within a minute to several minutes.

Since the design concentration for any total flooding system is **on the order of 34%, care must be taken when specifying a total flooding CO_2 system at a given location** (EPA 2015).

Discharge Time Delay

To prevent injury to occupants, it is imperative to evacuate the area prior to the CO_2 discharge. This is accomplished during system design through the use of an audible and visual pre-discharge alarm. Detection actuates a pre-discharge alarm and a time delay device, and only after the time delay device has completed a predetermined time delay sequence does it activate the CO_2 discharge.

Several factors must be taken into consideration in determining an adequate discharge time delay: the number of occupants that must be evacuated from the area, number of exits, travel distance to the exits, and the nature of the combustibles.

Another factor to consider is the spread of fire. The pre-discharge alarm feature, although absolutely necessary, can easily become a fire protection deterrent as the time can allow combustion to gain headway before the extinguishing action of the CO_2 begins. This becomes a major consideration when the material protected is in the Class A category, since deep-seated fires can result from combustion in paper, wood, and cotton. Because of its low cooling capacity, deep-seated combustion is difficult to extinguish with CO_2.

Re-Ignition

The smothering action of carbon dioxide will quickly extinguish the flaming fire created by a burning flammable liquid, but to completely extinguish the smoldering or deep-seated fire may require maintaining the CO_2 concentration for a considerable period of time. Heat is still present to re-ignite the fire. For example, the flame stage when bulk paper is burning can be quickly extinguished with a CO_2 concentration as low as 0.125 pcf (pounds of CO_2 per cubic foot of area volume), but temperatures of 700–1000 °F can still be present below the surface. As the CO_2 dissipates, the oxygen level builds up and re-ignition (or flashback) of the combustibles will occur. There have been cases where an individual can no longer see any sign of combustion and opens a door to enter the area. The heavier-than-air CO_2 rapidly flows out the open door and oxygen laden air rushes in to replace the CO_2 causing, in some instances, a combustion explosion.

Choosing the Time Delay

There are many variables to be considered in determining the length of time required for maintenance of the CO_2 concentration, and also the carbon dioxide concentration necessary to accomplish complete extinguishment in a deep-seated combustion condition. In most situations, the carbon dioxide concentration varies from 30% to 50% maintained from 30 to 60 min. There have been instances where an atmosphere close to a 100% CO_2 concentration had to be maintained for several days or weeks. Further criteria for choosing the time delay will be discussed in the carbon dioxide system criteria section.

Extended Discharge

An example of an extended discharge local application system is in the protection of a turbine generator used to produce electricity. These are massive machines that operate at a tremendous number of revolutions per minute, and the rotating shaft is lubricated by lubricating oil, piped and applied under considerable pressure. When a turbine generator is shut down, it continues to rotate for 30–45 min before coming to a complete stop, and during this run-down time, the shaft must be continually lubricated as the bearing-to-shaft tolerances are very close. The majority of turbine generators are operated by steam power, which, coupled with the high operating speed, creates a tremendous heating of the metal, usu-

ally far in excess of the ignition temperature of the lubricating oil. A rupture in an oil line, or a loosened connection in an oil line fitting caused by vibration, will spew oil onto the hot metal, and ignition will usually follow. An atomized oil spray produces an extremely hot fire that is very difficult to extinguish. CO_2 applied immediately will smother the flame, and the turbine generator unit can be shut down immediately, but lubricating oil must be continually pumped to the shaft during the run-down period or the bearings will bind. As soon as the CO_2 atmosphere clears, the oil will re-ignite on the hot metal. Enough CO_2 must be furnished to provide an extended, continuous or intermittent, discharge to maintain an inert atmosphere during the entire run-down period.

In evaluating the protection of turbine generators or other rotating equipment, one should note that metal, when heated, expands, and conversely when cooled, contracts. Size and shape are distorted by radical temperature changes. This distortion can be controlled to perform useful functions, but uncontrolled it can be destructive, especially with moving metal parts operating in close tolerance, as with the turbine generator between the rotating shaft and the bearings. A rapid, uncontrolled, sudden cooling of the rotating metal shaft or the bearings could create enough distortion to bind the moving parts and cause serious damage to the equipment. Placing the CO_2 nozzles too close to heated parts in motion can initiate a rapid lowering of the temperature when the CO_2 is discharged, and for this reason some turbine-generator CO_2 local application protection systems are designed to vaporize the CO_2 liquid through an electric or steam vaporizer prior to its discharge on the bearings.

Factory Mutual Insurance Company recommends the use of water spray for protection of turbine generator bearings based on results of their research which provided no evidence of metal distortion. One of the major manufacturers of turbine generators states that water spray protection will definitely cause metal distortion and, therefore, it is not to be used as a fire protection system agent of their equipment. This paradox is presented to accent the importance of treating each fire protection area carefully, basing system selection on the sum total of all hazard conditions.

Other Considerations

One design factor to keep in mind is that if the CO_2 protection is a total flooding system to protect a diesel or gasoline engine, the engine should be receiving its combustion air from outside of the room. If the CO_2 is sucked into the engine, there is the possibility of oxygen starved combustion and engine shut-down.

Warning signs are critical. Any room or area protected with a total flooding CO_2 system must be provided with warning signs to prevent personnel from entering the area following discharge, and to ensure that occupants are evacuated from the area before the system is discharged.

Carbon Dioxide and System Criteria

The extinguishing capability of all carbon dioxide systems depends on the method of discharge into the fire area, the rate of application, and the total quantity of carbon dioxide discharged. The method of discharge is determined by the nozzles selected and their location; the rate of application is determined by engineering calculations that establish the pipe and nozzle sizing, and the total quantity of carbon dioxide discharged depends on the calculated demand and the storage capacities.

There are two methods of storing carbon dioxide which characterize the system as either a high pressure or low pressure CO_2 system. In general, choice of a system requires an evaluation of whether to use 50–100 pound high pressure cylinders requiring considerable storage space, and some very substantial maintenance for periodic weighing of the cylinders to determine their full capability, and maintenance time to replace the cylinders that have discharged their CO_2, or to use a small, 1500 pound, low pressure tank.

High Pressure Systems

High pressure cylinders for CO_2 storage fixed pipe engineered systems are available in 50 pound, 75 pound, and 100 pound sizes. In considering the proper size of a cylinder to satisfy the required demand of a specific system, both the storage space available and the gross weight of the full cylinders are of importance. The use of 100 pound cylinders will reduce the required storage space, but because of the gross weight of the 100 pound cylinder, difficulty in handling becomes a factor. A 100 pound cylinder fully charged has a gross weight of about 300 pounds, while a 75 pound cylinder fully charged has a gross weight of about 215 pounds, and a 50 pound cylinder fully charged has a gross weight of about 165 pounds.

Carbon dioxide supply for a high pressure system is contained in high pressure cylinders which meet the test standards established by the ICC (Interstate Commerce Commission). High pressure systems operate on pneumatic principles-pressure is used to open valves, release doors and dampers, and to operate switches and accessories. Operating pressures are multiplied by large pistons. Operating parts are located internally.

The liquid carbon dioxide is stored in the cylinders at ambient or room temperature, and since ambient temperature determines the pressure in the cylinder, the container must be capable of withstanding the maximum pressure that could be expected. The recommended cylinder storage temperature range is 32–120 °F, with 130 °F being the maximum permissible temperature. Each cylinder has a frangible disc as a safety valve with a bursting pressure of 3000 psi, and at a maximum allowable temperature of 130 °F, there is still a margin of safety below this 3000 psi pressure point.

There are two important factors to note when choosing high pressure tanks: (1) temperature and its effects on CO_2, and (2) tank capacity. Basically, the temperature of the CO_2 affects its state-liquid, gaseous solid, or vapor. In order to regulate the system and maintain the proper state for fire protection, the temperature must be controlled. In addition, temperature affects the volume of the liquid expansion in the tank. When filling high pressure CO_2 cylinders, a vapor space is necessary above the liquid CO_2 in the cylinder to allow for expansion of the liquid due to temperature changes.

The following explanation attempts to describe the basic effects of temperature on the state (and tank capacity) of the cylinders in high pressure systems. When the storage temperatures of high pressure CO_2 cylinders drop below 32 °F, pressure decreases, and the rate of discharge can fall below design demand. If cylinder storage temperatures of below 32 °F are anticipated, a method of compensating for the resulting pressure decrease must be provided by heating the area of cylinder storage. Cylinders must be maintained at a minimum of 32 °F.

What happens when temperatures increase? As the temperature of liquid carbon dioxide in the cylinder increases, the CO_2 vapor pressure increases. As the CO_2 vapor pressure increases, the density of the CO_2 vapor over the liquid CO_2 increases; the liquid CO_2 expands and its density decreases and turns to a gas. At 87.8 °F, the critical temperature, the liquid and vapor have the same identical densities, and the liquid state has disappeared. Above 87.7 °F, CO_2 is entirely a gas regardless of the pressure.

Pressure affects the state and temperature of CO_2. At pressures below 60 psi, CO_2 reaches a point called the triple point. When stored at pressures below 60 psi, carbon dioxide can be present in a solid, liquid, or vapor state. Below the triple point, carbon dioxide can only exist as a solid or a gas, depending on the temperature. At 60 psi, the liquid carbon dioxide that remains in the cylinder or container will convert to dry ice at a temperature of −69 °F. If the pressure is reduced to atmospheric, about 15 psi, the temperature of the dry ice will drop to −110 F. When liquid carbon dioxide is discharged from the nozzles at atmospheric pressure, this same condition is established. When discharging CO_2 to extinguish a fire, the liquid converts to a vapor at approximately a 9: or 10:1 expansion ratio, and some of the vapor exists as dry ice particles at −110 F. The dry ice particles

will sublime or change to gas without passing through the liquid state. The extremely low temperature created will also cause condensation of water vapor in the atmosphere, creating a fog which sometimes lingers long after the disappearance of the dry ice particles.

At 70 °F, the pressure in a high pressure cylinder is 850 psi. Carbon dioxide expands as the temperature rises, but as long as an appropriate vapor space is maintained in the cylinder, the expansion does not increase the pressure. If the carbon dioxide cylinder was completely full of liquid CO_2, any rise in temperature, however slight, would rapidly increase the pressure in the cylinder. For this reason, the Department of Transportation (DOT) sets a limit on the amount of liquid CO_2 in a cylinder. The DOT criteria states that the weight of carbon dioxide liquid in a cylinder shall not exceed 68% of the weight of water that the container can hold at 60 °F. While this is a factor in specifying design criteria, tank manufacturers label their pressure capacities according to how much CO_2 the cylinder holds, rather than the tanks actual volume capacity. When a cylinder is listed as a 75 pound cylinder, it will contain 75 pounds of CO_2. The same is true of any cylinder capacity listing-if the description states that it is a 100 pound cylinder, there is a 100 pound CO_2 supply in the cylinder.

For example, if the hazard demand for CO_2 is 750 pounds of CO_2, the designer must select ten 75 pound high pressure cylinders. A reserve bank of cylinders may also be required (discussed later in this section), and the storage area for the cylinders will have to be large enough to accommodate twenty 75 pound cylinders, since the reserve supply must be equal to the primary supply.

Maintenance Considerations for High Pressure

The individual high pressure cylinders should be checked semi-annually for the required liquid content, which is done by weighing each cylinder. Most carbon dioxide equipment manufacturers provide a simple arrangement for weighing the cylinders without taking them out of service. The device consists of a very simple beam to which the individual cylinders are attached, and a direct reading scale. With a 10:1 ratio on the beam, a one pound pull would read 10 pounds.

When a cylinder weight has dropped by 10% of its required weight, it must be recharged, or even replaced if the loss is due to a pin-hole leak. Other common causes of loss through leakage of carbon dioxide from cylinders are corrosion and damage to the cylinders (due to mishandling). This points out the necessity to secure the cylinders in an upright position to eliminate any possibility of their being knocked over and selecting a safe storage area which must be outside the protected hazard area.

Cylinders are generally recharged at least at 5 year intervals, and they should also be hydrostatically tested. After a system has been installed for 12 years without being activated, the cylinders must be discharged, and each cylinder hydrostatically tested and refilled or replaced.

Reserve Cylinders

When a system with a demand for 750 pounds of carbon dioxide is activated by the detection system, manually or in the case of systems failure, the entire 10 cylinders will dump. The hazard area is then without protection until the cylinders can be removed and replaced with charged cylinders. This time lapse without fire protection, in most cases, should be eliminated with the installation of a reserve supply of carbon dioxide cylinders equal in volume to the primary supply. The reserve cylinders also provide a second discharge of CO_2 should the primary cylinder discharge not have controlled the fire condition.

Switching from the main to the reserve supply is a simple manual operation-the reserve cylinders are similar to the primary cylinders in piping and installation. A toggle switch usually redirects the signal from the main detection system to the reserve cylinders. As with the main cylinders, a manual pull station will activate the reserve cylin-

ders. There must be adequate visual indication that the remote station operates the reserve cylinder supply. The cylinder manifold (common pipe to which the cylinders are connected) serves the main and reserve cylinders. The reserve cylinders are isolated with full flow check valves to prevent the charged reserve cylinders from discharging into the empty cylinders. As a further precaution, each manifold has a bleeder valve which will automatically close when the manifold pressure exceeds 20 psi during discharge. Determination of the necessity for a reserve supply should follow a professional evaluation of the hazard relative to the frequency of fires, the areas need for uninterrupted operation, the availability of a human fire watch with hand extinguishing equipment and, most importantly, the speed with which the empty cylinders can be replaced with charged cylinders. In rare cases, it is possible to recommend eliminating the reserve supply, but this may be a penny-wise-and-pound-foolish decision, since fire ignition does not follow predictable and logical criteria.

How the System Works

To obtain an overview of how the system works, this section describes the process by which the CO_2 is propelled from the cylinder to the combustion. The pipes and fittings connecting the high pressure cylinders to the open nozzles are empty until the system is activated. When activated, the liquid carbon dioxide is transported to the nozzles through the piping. A siphon tube extends to the bottom of the cylinders so that liquid carbon dioxide will enter the piping. When this liquid CO_2 flows through the piping, there is a loss in pressure due to the increased volume of the empty piping and the friction between the moving liquid and the interior roughness of the pipe walls.

When the pressure of the liquid carbon dioxide drops, it changes to a vapor, and the piping fills with a mixture of liquid and vapor. As the mixture of liquid and vapor continues through the piping and the pressure continues to drop, the vapor increases in volume. This is the point where careful system design is critical. If the piping system is not designed properly, and the pressure reaches 75 psia (pounds per square inch absolute, which includes 15 psi atmosphere pressure) or 60 psi (the triple point) any further pressure loss would result in the remaining liquid flashing to a vapor and dry ice snow formation, which would clog not only the pipe but also the nozzle orifices.

Vaporization occurs as the first liquid carbon dioxide enters the piping and continues until an equilibrium is reached between the temperature of the pipe and the liquid carbon dioxide. The amount of liquid carbon dioxide vaporized, and the time required to reach this equilibrium with a resultant uniform flow, depend on:

- Flow rate
- Storage temperature
- Temperature, length, size, and piping material

It is necessary to adjust the total storage capacity and flow rate of the liquid carbon dioxide to compensate for vaporization and the time necessary to reach a flow equilibrium.

Design of the Piping System

The piping system must be designed to deliver the required rate of discharge through the selected nozzle orifices at the minimum nozzle pressure allowed, namely 300 psia for high pressure systems. This is the basic criteria for design of carbon dioxide piping systems (to be discussed further later in this chapter).

Once the required rate of discharge for the hazard is determined, and the piping and nozzles sketched, pipe sizes can be selected. The friction loss in fittings, valves, and cylinder connectors can be determined by changing the fitting or valve to an equivalent length of pipe. Tables are available to convert all sizes of fittings and valves to their equivalent pipe length, and charts will list the friction loss at various flows through various pipe sizes, with friction loss presented in psi per foot of pipe.

Terminal Pressures

Knowing the pressure at the CO_2 cylinders, and using the CO_2 manufacturers flow charts for friction loss, the terminal pressure at the nozzles can be calculated. If the terminal pressure at the nozzle is below 300 psia, it will be necessary to revise the pipe sizes to compensate for friction loss.

The piping for a high pressure carbon dioxide system may be subjected to extreme pressures. Thus, in order to resist the maximum pressures that could be expected, the pipe and fittings must have a minimum bursting pressure of 5000 psi. Valves that are under constant pressure must have a minimum bursting pressure of 6000 psi. The piping system must also be corrosion-resistant and be able to withstand the temperature extremes of a CO_2 system.

Discharge Nozzles

Carbon dioxide discharge nozzles come in a wide variety of designs and discharge patterns. There are basically two classifications: shielded, low velocity type, and the jet high velocity type. Each discharge nozzle consists of the orifice and usually a configuration of a horn, shield or baffle. The design of a discharge nozzle depends on whether it will be used on a total flooding or local application system.

The manufacturers nozzle orifice flow chart will indicate the proper nozzle orifice size required to deliver the desired flow rate at the nozzle pressure that was obtained from the piping calculations. A note of caution: If the manufacturers orifice flow chart indicates that the nearest available nozzle size is larger than required, it will be necessary to recalculate the piping system to provide the specified minimum nozzle pressure. Like the requirement for a piping system, carbon dioxide discharge nozzles should be of rugged construction to resist normal mechanical damage, extreme temperatures, and working pressures of the CO_2 system. The discharge orifice must be made of corrosion-resistant material. When nozzles are installed in areas where foreign material could clog the nozzle, they should be provided with frangible disc or blow-out caps that cover the nozzle orifice and are quickly ejected by the force of the discharging carbon dioxide.

Each discharge nozzle is permanently marked with the orifice size and is indicated as the equivalent diameter. This equivalent diameter refers to the orifice diameter of the standard single orifice type nozzle which has the same flow rate as the nozzle in question. The standard orifice is an orifice with a rounded entry with a coefficient of discharge not less than 0.98.

Thumbnail Calculation

It is possible to produce a layman's estimate of the amount of carbon dioxide required for a total flooding system. Remember that this thumbnail calculation is for preliminary discussions only. The actual engineering study, design, and calculations require a seasoned professional s expertise.

To get a quick estimate of how much CO_2 will be required for a total flooding system (high or low pressure system), determine the cubic footage of the area and multiply this figure by 0.10 pounds per cubic foot of CO_2. This thumbnail calculation does not take into consideration loss of CO_2 gas through enclosable openings and other factors that go into detailed calculations required for a CO_2 system. However, it will provide a preliminary demand of CO_2 for use in planning high pressure cylinder storage space, or the decision to use a low pressure storage tank instead.

High pressure CO_2 systems are more flexible than low pressure systems because selection of various sized cylinders −50, 75, and 100 pound - makes it possible to provide almost the exact amount of CO_2 required to meet the system demand. Low pressure systems are addressed below.

Low Pressure Carbon Dioxide Systems

When the total carbon dioxide demand is so large that it is not economical to store in multiple high pressure cylinders, a low pressure system is recommended. Massive banks of high pressure cylinders not only take up valuable storage space, but also require considerable maintenance time and labor, testing cylinder weight and replacing cylinders when necessary.

Low pressure carbon dioxide storage is a single refrigerated storage unit that is available in several storage capacities, ranging from 500 pounds to 50 tons. These storage tanks have their own built-in refrigeration unit which maintains the liquid carbon dioxide at 0 F, and a pressure of approximately 300 psi.

The low pressure liquid carbon dioxide storage tank is insulated, and refrigerant circulates through coils near the top of the tank. As the carbon dioxide vapor condenses at the coils, the tank pressure will vary. This tank pressure variance activates a pressure switch and automatically controls the compressor that circulates refrigerant through the coils.

The liquid carbon dioxide storage tank contains sufficient agent to supply the largest demand of a single hazard, or the individual demands of several hazards, plus a reserve supply of liquid CO_2. With the temperature controlled, filling density is not the serious consideration that high pressure cylinder storage presents. In addition, the low pressure unit has pressure relief valves to provide vaporization with subsequent cooling of the liquid, which prevents the storage tank from becoming liquid full. As a result, filling density for the low pressure storage unit is between 90% and 95%.

The Storage Tank and Refrigeration

The storage tank must be constructed, tested, equipped, and labeled in accordance with the Code for Unfired Pressure Vessels for Petroleum Liquids and Gases Specifications of the American Society of Mechanical Engineers (ASME). It must be equipped with a liquid level gauge, pressure gauge, and a high-low pressure supervisory alarm which activates an alarm if:

- the pressure in the tank rises to approximately 315 psi, or
- the pressure drops to 250 psi.

The refrigeration system is automatically controlled to maintain 0 °F in the storage tank under conditions where the highest anticipated tank storage area temperature is attained. If the low pressure storage unit is installed in an area where the temperature is expected to drop below 0 F, an automatic heating unit will be installed, capable of maintaining 0 °F in the pressure tank.

Should the refrigeration unit fail, and the pressure start to rise, relief valves on the unit are set to prevent the pressure from exceeding safe limits. When the pressure exceeds 341 psi, a diaphragm valve bleeds off the excess pressure, and if the pressure should continue to rise, a pop-off valve will operate at 357 psi, allowing a more rapid release of pressure. A final safeguard is a frangible disc which is designed to burst at 600 psi and vent the vapor to the atmosphere.

Where reliability of the refrigeration unit is a serious concern, dual refrigeration equipment is provided. The main concern for the reliability of refrigeration equipment is replacement of parts and down time required should the refrigeration unit need repair. Low pressure carbon dioxide storage tanks are so well insulated that even without the mechanical refrigeration, a four ton storage unit will not lose any carbon dioxide for at least 24 h, and at an 80° F outside air temperature, this size unit will only lose approximately 100 pounds per day.

Pipes and Fittings

The pipe, fittings, and valves supplying carbon dioxide from the low pressure storage unit to the nozzles must have a minimum bursting pressure of 1800 psi, compared with the minimum bursting pressure of 5000 psi for pipe and fittings and 6000 psi for valves in a high pressure carbon dioxide distribution system.

How the System Works

A low pressure carbon dioxide system is usually provided with a primary supply of carbon dioxide

to protect the major hazard, several smaller hazards, hose stations, and a reserve supply of equal volume. When a single low pressure unit supplies multiple systems, this is accomplished through the use of selector valves.

The master selector valve located at the storage unit automatically directs the CO_2 to satellite selector valves located in the immediate vicinity of each individual system. The master selector valve is activated when a satellite selector valve is opened by a signal from the specific hazard detection system or by manual operation of an individual hazard system through an electric/ manual pilot control system.

Dual Purpose Carbon Dioxide Units

An interesting use of the low pressure carbon dioxide system in a power generating facility is its dual use as a fire protection CO_2 supply and a CO_2 supply for use in purging the generators when they are emptied of the hydrogen atmosphere which is maintained internally when a generator is in operation. Hydrogen is commonly used as a cooling medium in large electric generators because of its very low wind resistance to the rotating equipment, and because it is a nonconductor of electric current.

Hydrogen is a highly flammable gas, and when the generator has to be worked on, it is imperative that all the hydrogen be evacuated. This can be accomplished by introducing CO_2 into the generator. The hydrogen is cleaned out or purged, and an inert atmosphere of CO_2 is created, nullifying the flammability of any hydrogen that might remain.

These dual purpose low pressure CO_2 units are sized so that the amount of CO_2 used for purging does not infringe on the supply of liquid CO_2 required for fire protection. A fire is possible in any number of protected hazards throughout the power generating facility, and especially during or immediately following the generator purging operation. By the use of shut off float valves, the supply of liquid carbon dioxide for purging cannot use any of the supply reserved for fire protection.

This same arrangement is used when the tank contains both the primary supply and the reserve supply of CO_2.

To prevent any chilling effect or dry ice snow when the CO_2 is introduced into the generator during the purging operation, or when CO_2 nozzles protect hot bearings, the liquid CO_2 is passed through an electric or steam vaporizer which converts the liquid CO_2 to a vapor. This vapor is then forced into the generator or discharged on the bearings entirely as a CO_2 gas.

Inerting an Area with Carbon Dioxide

Carbon dioxide is sometimes used as a fire prevention agent. Spontaneous heating and combustion in bunkers storing coal can present a costly fire condition that is very difficult to control and extinguish. Applying large volumes of water is the only reliable method, and it is rather obvious what large quantities of water can do to a coal supply, especially if it is finely ground coal.

Natural spontaneous combustion can occur in coal storage bunkers due to leakage of air at the bottom entrance gate of the bunker or through the top of the bunker. By introducing CO_2 gas through a number of injection points, it is possible to have the CO_2 vapor spread throughout and completely fill the bunker, even a bunker filled with finely ground coal. By maintaining a low continuous rate of application of CO_2 vapor (normally 50–100 pounds of CO_2 per hour), an effective inert atmosphere (insufficient oxygen to support combustion) can be maintained in the bunker. The air leakage through the bottom entrance gate can be reduced or even eliminated by the injection of carbon dioxide vapor into the hopper just above the gate.

Carbon Dioxide Hose Stations

The extinguishing capability of small hand-held CO_2 extinguishers is limited to small fires. The discharge range and duration of discharge of manual systems are quite limited-a few feet of range and a few seconds of discharge. The use of CO_2 hose stations will allow an operator to apply a large capacity of CO_2 from a safe distance away from the combustion.

Carbon dioxide hose stations can be supplied from high pressure CO_2 cylinders, or from a low pressure CO_2 tank. The hose reels are filled with CO_2 by an automatic or manual release. The manual release may be a pull station, push button, or similar device located adjacent to the hose reel and clearly identified by a permanent sign.

One familiar automatic device is the bracket that holds the extended length of hose and discharge nozzle, commonly called the discharge horn. When the discharge horn is removed from the bracket, the CO_2 is automatically released into the hose. The discharge horn nozzle is equipped with a squeeze-grip type, quick opening and closing valve, so that the operator has complete control of the CO_2 discharge.

The supply of CO_2 for a hose station shall be sufficient to supply the discharge rate for at least 1 min. If the hose station is to be used by an inexperienced person who might waste some CO_2, the supply should be increased. Multiple hose stations supplied from a single supply should have a supply adequate to accommodate all hose stations likely to be used at one time, at the determined discharge rate for 1 min.

The hose is a high pressure type with a bursting pressure of 5000 psi if the supply is from high pressure cylinders, and 1800 minimum bursting pressure if the supply is from a low pressure tank. Three-quarter inch hose will permit a 100–150 pound-per-minute discharge rate of CO_2, and by using a 1″ or larger hose, discharge rates of 250–300 pounds per minute with a discharge rate at 25′.

The location and number of hose stations should be determined by the equipment being protected. If possible, the hose reel should be placed close to an exit door so that if smoke and toxic fumes from the combustion become excessive, the person using the hose can quickly exit the area. Anyone using CO_2 hose stations should be properly trained in firefighting and safety precautions. CO_2 is 50% heavier than air and will flow like water to low points in the building such as pits, basements, tunnels, etc. Thus caution must be seriously considered when there has been a considerable amount of CO_2 discharged from a local application system and/or CO_2 hose stations.

The Detection System and Air Leakage

The detection system that activates automatic CO_2 total flooding systems provides the alarm, but always specify a pressure switch on the discharge piping of the CO_2 system to activate the alarms. If someone manually activates the CO_2 system by a mechanical device, the detection/alarm system will be bypassed and in all probability the CO_2 will be discharged without activating any alarms.

The pressure switch to activate the alarm system can also be used to shut down the ventilation system in a room protected with a total flooding CO_2 system. It may also release magnetic door holders that hold self-closing doors in the open position, allowing them to close automatically. The magnetic door holder releases its magnetic hold on the door when a signal is received from the pressure switch.

An engineering evaluation of the protection may mandate that the HVAC (heating, ventilating, and air-conditioning) system be shut down, dampers closed, and magnetic door holders released as a result of a signal from the detection system, so that the room or area will be as tight as possible when the CO_2 discharges. Operating from the pressure switch activated by the discharge pressure of the CO_2 may present a situation where the doors close too slowly, HVAC fans keep rotating after power is shut off, and too much CO_2 will be lost, effecting the rapid build-up of the required CO_2 concentration.

Carbon Dioxide System Design

After clarifying the various types of CO_2 systems low and high pressure, and total and local applications), we can now explain the details involved in designing these systems.

Local Application System Design

As previously described, local application carbon dioxide systems are used to protect against individual hazards, hazardous operations, processes, and equipment where it is impractical to enclose them and use a total flooding system. Total flooding may be impractical due to the low hazard conditions that exist in the surrounding area where the hazard is located, the number of people working in the area, the impossibility of shutting down machinery prior to leaving the area, or just the economics involved because of the size of the total area that would have to be total flooded with CO_2. However, the local application system does not have the advantage of walls that confine the CO_2 (as does the total flooding system), and the CO_2 discharge is rapidly dissipated. As a result, the design of local application hazard protection can be complex.

The local application system consists of CO_2 nozzles that discharge directly onto or into a specific hazard or hazard area. The hazard can be the surface of a flammable liquid or a piece of equipment. There are basically two methods of designing a local applications system: the rate-by-area method and the rate-by-volume method.

Rate-By-Area Method

When the hazard involves a flat surface or hazards basically oriented on a horizontal plane that can be readily measured, the rate-by-area method is used for design. Examples of such hazards are flammable liquid dip tanks and drain boards, quench tanks, open tanks, etc. Before calculating and designing a local application CO_2 system by the rate-by-area method, it must be clearly understood that each type of CO_2 discharge has an approval rating which indicates its individual design data. Design data consists of:

- Rate of discharge
- Spacing limitations
- Maximum area of coverage at specified heights

- Design flow rate versus height or distance from the hazard surface
- Projection angle versus distance and height from hazard

Below are the steps involved in designing a local application system using the rate-by-area method.

1. Extent of Hazard Area. The first step in using the rate-by-area design method is to carefully determine the extent of the hazard area including not only the hazard itself, but also fringe areas that may be involved and any discharge obstructions that could be present in or adjacent to the hazard.
2. Prepare a Layout. The second step is to prepare a layout, drawn to scale, of the hazard and any ramifications of the hazard which would effect the discharge patterns of the nozzles. With this layout, nozzles can now be spotted around and/or above the hazard. Tankside or linear-type nozzles, if applicable, can be strategically located to take full advantage of their approval listings as to discharge rate, coverage, and discharge pattern.
3. Area of Coverage. The use of overhead nozzles located above the hazard are restricted by the height and area limitations stipulated in the approval listings for each nozzle. The approval listings furnish information relative to area of coverage at various heights. The area of coverage from overhead nozzles is considered to be in the form of a square.
4. Height of Overhead Nozzles. The height of overhead nozzles above the hazard determines the optimum flow rate for extinguishment of the specific hazard material being protected. This relationship of height, optimum flow rate, and hazard material is given in the individual approved criteria of each nozzle.
5. Minimum Discharge Time. The minimum discharge time for the nozzles is set at 30 s, but the hazard material may dictate an extension of this minimum. Metal heated by the fire before extinguishment may demand anf extension of the discharge time in order to cool the metal below the ignition temperature of the

hazard material. All forced drafts, unusual air movement, and wind currents in the area of the hazard must be carefully evaluated and compensated for by additional nozzles or relocation of nozzles to protect areas outside the hazard areas.

6. Total Nozzle Flow Rate. Once the discharge duration time is established, and the flow rates of the individual nozzles are determined, the total carbon dioxide supply required becomes as a simple multiplication problem of total nozzle flow rate demand times duration of discharge.
7. Sizing and Routing of Piping. The number of cylinders has now been determined, and a location for the cylinders remote from the hazard area is established. All that remains is sizing and routing of the piping from the cylinders to the nozzles. This phase of the project is the most complicated, and the part requiring the most accurate, exacting, and precise computations. Carbon dioxide equipment manufacturers prepare tables, charts, and graphs which are a help in these calculations, but the expertise of an engineer experienced in CO_2 systems is required in all cases except the simplest project.

The general criteria for determining the placement of overhead nozzles for a rate-by-area design is as follows:

1. Distance between nozzles shall not be more than the square root of the area rating.
2. Distance to the edge of a hazard shall not be more than half the distance calculated in (1) above.
3. If possible, overhead nozzles should be installed over the center of the hazard and perpendicular to the hazard surface.
4. Overhead nozzles installed at an angle to the hazard shall not be installed at an angle less than 45 from the surface plane of the hazard.
5. The distance from the face of the overhead nozzle to the aiming point of the nozzle on the hazard surface shall be used to determine the height with reference to determining required flow rate and area coverage.

6. When installing overhead nozzles on an angle, use NFPA 12 to determine the distance of the aiming point from the near side of the surface of the hazard by multiplying the width of the nozzle coverage by the appropriate aiming factor.

Rate-By-Volume Method

The rate-by-volume design method of local application is used when:

(1) the hazard is three-dimensional or an irregular physical configuration that cannot be accurately measured and reduced to equivalent surface areas, or
(2) hazards are partially enclosed but have openings that are too large to consider a total flooding system. Examples of such hazards are transformers, printing presses, and rolling mills. The steps involved in designing a local application system using the rate-by-volume method follow.

1. Assume an Enclosure. The first step is to assume an enclosure around the hazard with the imaginary walls and ceiling a minimum of 2′ from the sides to the top of the hazard. Now that the hazard has been boxed-in, compute the volume enclosed in the imaginary box. This volume calculation is the total volume of the assumed enclosure with no deduction for the volume of any solid object within the enclosure. If the hazard is small, a minimum four square foot enclosure shall be used.
2. Set Minimum Discharge Times. Once the cubic footage of the assumed enclosure is determined, the standard total discharge rate of carbon dioxide is one pound per minute per cubic foot of assumed volume for a minimum of 30 s. This discharge rate of 1 lb/min./cu. ft. is National Fire Protection Association criteria, which differs from the Factory Mutual Insurance required discharge rate of 2 lb/min./cu. ft. As explained in the rate-by-area design section, this minimum discharge time may have to be increased

depending on the hazard conditions, forced drafts, or unusual air currents. The assumed volume shall be increased to compensate for the loss of CO_2.

3. Permanent Walls. If there are permanent walls enclosing part of the hazard, and if these walls extend at least 2′ above the hazard, the total discharge rate of one pound of CO_2/minute/cu. ft. may be reduced proportionately, but never below 0.26 pounds/minute/cu. ft., and only after a careful evaluation of the hazard material and conditions. Any permanent wall which forms a partial enclosure must cover one entire side of the imaginary enclosure, and the length of this wall must be at least 25% of the perimeter of the enclosure. If one wall complies with this criteria, and there are adjoining walls that comply with the height requirement but do not extend the full length of the hazard area, these short walls may be included in calculating the percentage of the perimeter that is enclosed.

Walls that meet the length and height criteria but are opposite each other and therefore not adjoining can only be considered as one wall. If the two opposite wall lengths added together equal at least 50% of the imaginary hazard enclosure perimeter, their total length may be used in calculating the required rate of discharge.

The discharge rates of these semi-enclosures for calculations based on the rate-by-volume method are calculated using prepared tables found in NFPA standards.

If the base of the semi-enclosed or imaginary hazard enclosure has openings that exceed 3% of the total base area, the rate-by-area method of calculations must be used to compute the discharge rate. If the openings in the base or floor are 3% or less of the total base or floor area, 10 pounds per minute per square foot of opening additional CO_2 must be included in the discharge rate calculations. If a permanent wall forming a partial enclosure of the hazard has openings which total more than 10% of the wall area, the wall cannot be considered as part of the enclosure.

If the total wall openings equal 10% or less of the total wall area, two pounds per minute per square foot of opening additional CO_2 must be included in the discharge rate calculations.

Special Considerations

Dip tanks and drainboards may have coated parts that hang on a conveyor above the flammable liquid surface. In this hazard consideration, the flammable liquid surface CO_2 protection will be figured using the rate-by-area method. If the coated parts extend more than 2 above the flammable liquid surface, this area must be protected by computing the discharge rate using the rate-by-volume method. An imaginary enclosure is again used, assuming the base to cover the same area as the protected surface and extending from 2 above this surface to the maximum height of the coated parts, and calculating two pounds of CO_2 per minute per cubic foot of the total volume of this assumed enclosure as a rate of discharge. Hoods over kitchen ranges and hoods over special processes requiring exhaust systems (to collect and exhaust fumes so they do not permeate the room atmosphere) must be protected if the cooking facilities involve flammable grease or fat, or if the special processes involve flammable liquids.

If the area below the hood is protected with carbon dioxide (such as deep fat fryers, process burners, or liquid heaters), the rate of discharge in the hood is figured as 0.6 pounds of CO_2 per minute per cubic foot of hood volume. If the area below the hood is not protected, the rate-by-area method is used to calculate the discharge rate, using the entire interior surface of the hood.

Total Flooding System Design

Total flooding system design requires a thorough knowledge of the hazard, the enclosure, type of fire expected, enclosure leakage, and ventilation. Calculations can be quite accurate because the volume need not be assumed, but can be calculated, and the loss of carbon dioxide from the enclosure is a known factor.

The design and calculations for a total flooding system must be adequate to provide the required concentration within 1 min of the start of the discharge into the room or area for surface fire conditions. For deep-seated fires, the design concentration shall be completed within 7 min, with a rate of discharge to provide a 30% concentration within 2 min. The temperature of the area to be protected with a total flooding CO_2 system will effect the fire suppression efficiency of carbon dioxide. To compensate for relatively high temperature areas, an additional 1% of carbon dioxide is required for every 5° over 200 °F in the temperature of the area being protected. Following are the steps involved in designing a total flooding system.

1. Percent of CO_2 by Volume. Charts published by the NFPA and carbon dioxide manufacturers indicate the various carbon dioxide concentrations, or percent of CO_2 by volume, required to extinguish common liquid and gas surface fires by total flooding. It will be noted that these charts indicate Theoretical and Minimum Design concentrations. The reason for this is the fact that some allowance is needed to compensate for the lack of complete and perfect diffusion of CO_2 throughout the entire area being total flooded at any point in time during the inerting period.

2. Tight Enclosure. As mentioned previously, to attach effective total flooding, the area must be a relatively tight enclosure. HVAC (heating, ventilation, and air-conditioning) systems must be shut down automatically before the CO_2 discharge commences. If this cannot be accomplished, additional carbon dioxide must be provided to compensate for the CO_2 lost through the HVAC system (due to HVAC fan run-down after being shut off, thereby causing a loss of CO_2).

3. Stoppage of Flammable Materials. In addition, flammable liquid pumps, conveyors, and mixers must also be stopped to discontinue the flow of combustible material into the area. Windows, dampers, and doors are shut automatically with pneumatic release devices actuated by CO_2 piping discharge pressure.

The pressure in the CO_2 piping when discharge commences enters the pneumatic release device, and this pressure performs mechanical work which results in an electrical contact; electrical power is supplied to alarms, open dampers, and door closers.

4. Rate of Discharge. The total flooding rate of discharge for surface fires must be ample to reach the minimum discharge concentration within 1 min after start of discharge. Flow equilibrium (liquid-to-vapor) in the piping with high pressure systems is usually an insignificant time element, and therefore the rate of discharge will be the total required volume divided by 1 min. Low pressure systems can require a significant amount of time and CO_2 quantity in vaporization and cooling of the piping before a flow equilibrium is reached to demand a calculation to increase the equilibrium flow rate so that the recommended total quantity will be delivered within the one-minute-after-start of discharge requirement. In other words, delay time and weight of CO_2 vaporized during initial discharge must be included in the rate of discharge calculation.

5. Time Delay of Vaporization. Using the tables and formulas found in NFPA Standards and Factory Mutual Insurance data sheets, calculations will determine the time delay of vaporization in the piping and weight of CO_2 required to overcome the loss through vaporization in the piping.

6. Concentration Level for Deep-Seated Fires. In order to extinguish the deep-seated or smoldering fire, the total flooding system must provide a 30% concentration within 2 min from start of discharge, and this rate of discharge can be computed on the basis of 0.042 pounds/cu. feet. As mentioned previously in the discussion of total flooding systems, when the required design concentration is reached, surface fires will be extinguished, but with deep-seated or smoldering fires it is necessary to maintain the concentration for an extended period of time in order to soak the material with CO_2 until it reaches the deepest penetration of the combustion in the material. Thirty minutes is the average holding time for closely packed combustible material and occupancies such as fur vaults and record storage.

Calculations for the concentrations necessary to extinguish deep-seated combustion are difficult to establish, as the concentration required varies with the mass and physical arrangement of the material. Dense mass material has a high thermal insulating effect which permits combustion to retain the heat produced and prevents rapid soaking of the mass with the CO_2 gas. Tests have been conducted using the most common materials found in record storage vaults, dust collectors, fur storage vaults, cable rooms where insulated wiring is found, and small electric motors and machines, to establish the criteria to be used in designing total flooding systems for these hazards. As a result of these tests, it has been established that design concentrations of between 50% and 75% and flooding factors of between 0.083 and 0.166 are used in calculating the quantity of carbon dioxide required for each specific hazard listed above. Examples: electric equipment and wiring, 50% design concentration, 0.083 pounds of CO_2 per cubic foot of volume. For storage vaults, 75% design concentration, 0. 166 pounds of CO_2 per cubic foot of volume.

7. Quantity of CO_2 To calculate the quantity of CO_2 required to extinguish a surface fire by total flooding in an enclosure, the following are the volume factors required for a few common materials.

Material	Minimum concentration of CO_2%
Benzene	37
Ethyl ether	46
Gasolene	34
Kerosene	34
Lube oil	34

Do not neglect openings in the enclosure that cannot be automatically closed at the start of discharge of the CO_2 The loss of CO_2 through these openings must be included in the total supply calculations at not less than one pound of carbon dioxide per square foot of openings. HVAC systems that cannot be shut off automatically at the start of discharge of the CO_2, or cannot be immediately shut off manually, must be given careful consideration in computing the CO_2 supply. The volume moved by the operating ventilation, or HVAC, system during the time span between CO_2 initial discharge into the area and mechanical unit shut-down should be divided by the flooding factor, and this amount added to the total supply requirement. When the design concentration is greater than 34%, this calculated amount must be multiplied by a conversion factor found in NFPA standards.

8. Unusual Temperature Conditions. An additional design consideration is the existence of an unusual temperature condition in the enclosure to be flooded. For every 5° above 200 °F the total calculated quantity of CO_2 must be increased by 1%. When the normal temperature of the enclosure is below 0 °F, the total calculated quantity of CO_2 must be increased by 1% for every degree below 0 °F.

General Design Notes

A few design notes to supplement the general discussion of pipe and fittings for high and low pressure systems. General:

1. Pipe and fittings must be suitable for low-temperature use.
2. Pipe and fittings must be corrosion resistant inside and out.
3. Ferrous metals shall be galvanized.
4. Steel, copper, brass and metals with similar mechanical and physical properties are acceptable.
5. Copper tubing with approved flared or brazed connections is acceptable.
6. Malleable or ductile iron fittings are acceptable if they meet ASTM (American Society of Testing Material) A395, Grade 60-45-15.
7. Steel pipe that meets ASTM A-53 has excellent low-temperature characteristics.
8. Cast (gray) iron pipe and fittings are not acceptable.

9. Where pipe sections can be closed off, pressure-relief valves shall be installed to vent entrapped liquid carbon dioxide.
   (a) High pressure systems: pressure relief set to operate at 2400–3000 psi.
   (b) Low pressure systems: pressure relief set to operate at 450 psi.
   (c) Care must be exercised to discharge CO_2 from pressure-relief valves without causing personal injury or mechanical damage.
10. The area required for venting of exceptionally tight enclosures can be calculated with the following formula. This formula is based on the assumption that expansion of carbon dioxide is nine cubic feet per pound:

X equals Q divided by 1.3 times the square foot of P.

X = Free venting area in square inches.

Q = Calculated carbon dioxide flow rate in pounds per minute.

P = Allowable strength of enclosure in pounds per square foot.

Note: See NFPA 12 for a guide to substituting a value for P (Strength and allowable pressure for avg. enclosures). These values are based on general construction practice parameters.

This section has provided an overview of a CO_2 system-high vs. low pressure, and local vs. total flooding applications. This data can be used to generally make a choice between systems, however, in choosing the valves, pipes and design details, an appropriate engineer should be consulted.

Clean Agents-Replacing Halon

The Montreal Protocol of 1987 on Substances that Deplete the Ozone Layer resulted in the phase out of a number of different chemicals, including chlorofluorocarbons (CFCs) and Halons.

The result was a concerted effort to develop halon substitutes. Three common substitutes are Inergen, FM-200, and Novec, although there are others. These gases are known as clean agents. They do not conduct electricity, vaporize readily and theoretically leave no residue. There are two basic types -Inert agents and halocarbon clean agents.

It is important to note that by their nature the Halon substitutes will occupy more occupancy space than Halon systems.

Types of Clean Agents

To paraphrase the SFPE Handbook (Hurley et al. 2015), halocarbon clean agents are:

- All electrically nonconductive.
- They all vaporize readily and leave no residue.
- All are liquefied gases or display similar behavior.
- All (except FE30) use nitrogen super-pressurization for discharge purposes in most cases.
- All by chemical composition are LESS efficient extinguishing agents than halon.
- All are total flooding gases after discharge.
- All produce more hydrofloric acid and other decomposition than Halon 1301.
- Most have substantial greenhouse warming characteristics.
- All have near zero ozone depletion potential.
- All must be evaluated with respect to health and safety concerns -primarily cardiac sensitization.

Inert clean gas agents:
All are electrically nonconductive.
- All are clean agents; they leave no residue.
- All are stored as compressed gases using low capacity high pressure cylinders (and therefore can potentially take up a lot of space)
- All are less efficient fire extinguishants than Halon 1301. Therefore storage volumes are much greater than Halon 1301 or the halocarbon agents mentioned previously.
- Don't decompose -that is they do not produce HF or other products of decomposition
- Have zero global warming potential.
- Have zero global ozone depletion potential.

Must be evaluated with respect to health and safety concerns which are primarily related to oxygen depletion.

Greenhouse Gases and the Ozone Layer

Ozone Depletion Potential

The Earth's atmosphere is composed of several layers. The stratosphere is located 6 miles (10 km) to 31 miles (50 km) above the earth's surface. The majority of the ozone layer is located in the stratosphere. Increases in the ozone layer near the earth's surface are the result of pollution from human activities (smog).

The ozone layer absorbs a portion of the radiation from the sun and prevents it from reaching the planet's surface. Its most important job is to absorb a portion of UV light called UVB, which has been linked to many harmful effects, including cataracts, skin cancers and harm to some crops and marine life.

Ozone consists of three oxygen molecules and has the chemical equation O_3. When chlorine and bromine (which are contained in Halons) come into contact with the stratosphere, they react with the O_3 molecules, destroying them. Hence replacement for the halons was deemed necessary by the Montreal Protocol on Substances that Deplete the Ozone Layer was agreed on September 16, 1987 (EPA, OLP 2017).

Global Warming Potential

Greenhouse gases (GHGs) warm the Earth by absorbing energy and slowing the rate at which the energy escapes to space; they act like a blanket insulating the Earth. Different GHGs can have different effects on the Earth's warming. Two key ways in which these gases differ from each other are their ability to absorb energy (their "radiative efficiency"), and how long they stay in the atmosphere (also known as their "lifetime").

The Global Warming Potential (GWP) was developed to allow comparisons of the global warming impacts of different gases. Specifically, it is a measure of how much energy the emissions of 1 ton of a gas will absorb over a given period of time, relative to the emissions of 1 ton of carbon dioxide (CO_2). The larger the GWP, the more that a given gas warms the Earth compared to CO_2 over that time period. The time period usually used for GWPs is 100 years. GWPs provide a common unit of measure, which allows analysts to add up emissions estimates of different gases (e.g., to compile a national GHG inventory), and allows policymakers to compare emissions reduction opportunities across sectors and gases (EPA, E 2017)

Naming Conventions

It should be noted that clean agents are referred to in abbreviated form in the NFPA 2001 (2015) standard. This is shown in Table 12.1.

Inert Agents

A common inert gas used to suppress fires is Inergen. It literally inerts an environment -an inert chemical mixture drops the oxygen level to a point where combustion can no longer be sus-

Table 12.1 Trade names, chemical composition, and NFPA references

Trade name	Chemical composition	NFPA reference
Inergen	Nitrogen (52%) Argon (40%) Carbon Dioxide (8%)	IG-541
FM-200	CF3CHFCF3	HFC-227ea
Novec	C6F12O	FK-5-12

Reprinted with permission from NFPA 2001-2018, *Standard on Clean Agent Fire Extinguishing Systems,* Copyright © 2017, National Fire Protection Association, Quincy, MA. This reprinted material is not the complete and official position of the NFPA on the referenced subject, which is represented only by the standard in its entirety which may be obtained through the NFPA website at www.nfpa.org

tained. A common composition is derived from a mixture of Nitrogen (52%), Argon (40%), and Carbon Dioxide (8%). Inergen is suitable for Class A, Class B and Class C fires. It has zero ozone depletion potential, is electrically non-conductive and is safe for occupied areas. One drawback is that it generally requires many more cylinders and therefore more space, than the other ozone friendly gases, since its constituent gases to not lend themselves to compression without refrigeration at sea level.

The main life safety concern for inert agents like Inergen is reduced oxygen concentration.

Halocarbon Agents

These agents have a drawback that when used on larger design fires, it is possible that Hydrogen Fluoride - a very strong acid, is produced in the area of the fire. This can be mitigated by using special detection systems in the hazards to be protected, so that smaller fires are detected and initiate the suppression system (limiting HF production).

Besides the development of HF, cardiac sensitization leading to irregular heartbeat (and possible heart attack) is a primary toxicity problem for the Halocarbon agents at higher concentrations, so much care must be taken when designing these new systems.

One halocarbon agent is FM-200. FM-200 is a carbon/fluorine/hydrogen compound suitable for Class A, B and C fires. It removes heat so that the combustion reaction cannot sustain itself. It also has Zero ozone depletion potential, is electrically non-conductive and is safe for occupied areas.

The base chemical in FM-200 is used in medical inhalers. While this can provide some relief to those who experience an accidental discharge it should not necessarily comfort those who are exposed to the gas when it has been broken down at high temperatures.

Another is Novec. Novec is a suitable for Class A, B and C fires. It removes heat so that the combustion reaction cannot sustain itself. It also has a very low ozone depletion potential, is electrically non-conductive and is safe for occupied areas.

NOAEL and LOAEL and the Need for Precise Design

NOAEL is a measure of agent toxicity to humans. It is an acronym for No Observed Adverse Effect Level. LOAEL is also an acronym for Lowest Observable Adverse Effect Level (on test subjects).

> **1.5.1.2.1*** Unnecessary exposure to halocarbon clean agents — including exposure at and below the no observable adverse effects level (NOAEL) — and halocarbon decomposition products shall be avoided. Means shall be provided to limit exposure to no longer than 5 min. Unprotected personnel shall not enter a protected space during or after agent discharge. The following additional provisions shall apply:
> (1) Halocarbon systems for spaces that are normally occupied and designed to concentrations up to the NOAEL shall be permitted. The maximum exposure in any case shall not exceed 5 min.
> NFPA 2001(2015).

Needless to say, clean agent designers must use methods to minimize human exposure to the agent. The best way to do this is to make sure that exiting from a compartment that utilizes such a system is well facilitated. Other methods to protect personnel include alarm notification before discharge, locks and other methods to prevent re-entry into the room once a system has been discharged.

Quantities of Clean Agents

In many applications, the number of cylinders and general "footprint" of a system become very important. In cities where space is at a premium, or offshore where weight and space may be constraints, it is important to have a sense of how these factors will influence a design.

For very large spaces of 17,660 ft³ (500 m³) and more, comparisons were made of the sizes and weights of different types of systems (Wickham 2003). These are summarized below:

Agent	Cylinder volume (Liters)	# of cylinders	Total weight (kg)
Halon 1301	246	1	400
Carbon dioxide	68	8	1000
FM-200 (halocarbon)	368	1	600
Novec 1230 (halocarbon)	368	1	600
Inergen (inert gas)	82	19	2000

The number of cylinders and weights make it clear that Inergen and CO_2 may be undesirable from a weight and pure system volume standpoint depending on the area that is to be protected.

For a smaller space, approximately 2500 cubic feet (72.2 m³), an aerosol fire suppression vendor made the following comparisons for weights and numbers of cylinders: (Stat-X 2005). Aerosol fire suppression systems are covered under NFPA 2010. Health issues concerning breathing the products of aerosol fire suppression systems and the obscuration of exits that may occur upon deployment mean that they should only be employed in conjuction with a where it can be assured that no occupants will be present (a 30 s time delay is required).

Agent	# of cylinders	Total weight (kg)
Aerosol fire suppression	2	5
Dry chemical	2	22.7
Halocarbon	2	59
Carbon dioxide	4	73
Inert gas	9	52

Quantities of clean agents can be determined using the tables provided in NFPA 2001 (2015). For examples for Inergen of these tables are provided in Tables 12.2 and 12.3.

Table 12.2 Total Flooding Quantity of Inergen

2001–84   CLEAN AGENT FIRE EXTINGUISHING SYSTEMS

Table A.5.5.2(e) IG-541 Total Flooding Quantity (U.S. Units)[a]

Temp (t) (°F)[c]	Specific Vapor Volume (s) (ft³/lb)[d]	Volume Requirements of Agent per Unit Volume of Hazard ($V_{agent}/V_{enclosure}$)[b] Design Concentration (% by Volume)[e]							
		34	38	42	46	50	54	58	62
-40	9.001	0.524	0.603	0.686	0.802	0.873	0.977	1.096	1.218
-30	9.215	0.513	0.590	0.672	0.760	0.855	0.958	1.070	1.194
-20	9.429	0.501	0.576	0.657	0.743	0.836	0.936	1.046	1.166
-10	9.644	0.490	0.563	0.642	0.726	0.817	0.915	1.022	1.140
0	9.858	0.479	0.551	0.628	0.710	0.799	0.895	1.000	1.116
10	10.072	0.469	0.539	0.615	0.695	0.782	0.876	0.979	1.092
20	10.286	0.459	0.528	0.602	0.681	0.766	0.858	0.958	1.069
30	10.501	0.450	0.517	0.590	0.667	0.750	0.840	0.939	1.047
40	10.715	0.441	0.507	0.578	0.653	0.735	0.824	0.920	1.026
50	10.929	0.432	0.497	0.566	0.641	0.721	0.807	0.902	1.006
60	11.144	0.424	0.487	0.555	0.628	0.707	0.792	0.885	0.987
70	11.358	0.416	0.478	0.545	0.616	0.693	0.777	0.868	0.968
80	11.572	0.408	0.469	0.535	0.605	0.681	0.762	0.852	0.950
90	11.787	0.401	0.461	0.525	0.594	0.668	0.749	0.836	0.933
100	12.001	0.393	0.453	0.516	0.583	0.656	0.735	0.821	0.916
110	12.215	0.386	0.445	0.507	0.573	0.645	0.722	0.807	0.900
120	12.429	0.380	0.437	0.498	0.563	0.634	0.710	0.793	0.884
130	12.644	0.373	0.430	0.489	0.554	0.623	0.698	0.779	0.869
140	12.858	0.367	0.422	0.481	0.544	0.612	0.686	0.766	0.855
150	13.072	0.361	0.415	0.473	0.535	0.602	0.675	0.754	0.841
160	13.287	0.355	0.409	0.466	0.527	0.593	0.664	0.742	0.827
170	13.501	0.350	0.402	0.458	0.518	0.583	0.653	0.730	0.814
180	13.715	0.344	0.396	0.451	0.510	0.574	0.643	0.718	0.801
190	13.930	0.339	0.390	0.444	0.502	0.565	0.633	0.707	0.789
200	14.144	0.334	0.384	0.437	0.495	0.557	0.624	0.697	0.777

[a]The manufacturer's listing specifies the temperature range for operation.

[b]X [agent volume requirements (ft³/ft³)] = volume of agent required per cubic foot of protected volume to produce indicated concentration at temperature specified.

$$X = 2.303 \times \left(\frac{s_0}{s} \right) \times \log_{10}\left(\frac{100}{100-C} \right) = \left(\frac{s_0}{s} \right) \times \ln\left(\frac{100}{100-C} \right)$$

where:

s_0 [specific volume (ft³/lb)] = specific volume of inert gas agent at 70°F and 14.7 psi absolute

[c]t [temperature (°F)] = design temperature in the hazard area.

[d]s [specific volume (ft³/lb)] = specific volume of IG-541 vapor can be approximated by $s = 9.8579 + 0.02143t$, where t = temperature (°F).

[e]C [concentration (%)] = volumetric concentration of IG-541 in air at the temperature indicated.

Reprinted with permission from NFPA 2001-2018, *Standard on Clean Agent Fire Extinguishing Systems*, Copyright © 2017, National Fire Protection Association, Quincy, MA. This reprinted material is not the complete and official position of the NFPA on the referenced subject, which is represented only by the standard in its entirety which may be obtained through the NFPA website at www.nfpa.org

2015 Edition

Table 12.3 Total Flooding Quantity of FM200

2001–70   CLEAN AGENT FIRE EXTINGUISHING SYSTEMS

Table A.5.5.1(i) HFC-227ea Total Flooding Quantity (U.S. Units)[a]

Temp (t) (°F)[c]	Specific Vapor Volume (s) (ft³/lb)[d]	Weight Requirements of Hazard Volume, W/V (lb/ft³)[b] Design Concentration (% by Volume)[e]									
		6	7	8	9	10	11	12	13	14	15
10	1.9264	0.0331	0.0391	0.0451	0.0513	0.0570	0.0642	0.0708	0.0776	0.0845	0.0916
20	1.9736	0.0323	0.0381	0.0441	0.0501	0.0563	0.0626	0.0691	0.0757	0.0825	0.0894
30	2.0210	0.0316	0.0372	0.0430	0.0489	0.0550	0.0612	0.0675	0.0739	0.0805	0.0873
40	2.0678	0.0309	0.0364	0.0421	0.0478	0.0537	0.0598	0.0659	0.0723	0.0787	0.0853
50	2.1146	0.0302	0.0356	0.0411	0.0468	0.0525	0.0584	0.0645	0.0707	0.0770	0.0835
60	2.1612	0.0295	0.0348	0.0402	0.0458	0.0514	0.0572	0.0631	0.0691	0.0753	0.0817
70	2.2075	0.0289	0.0341	0.0394	0.0448	0.0503	0.0560	0.0618	0.0677	0.0737	0.0799
80	2.2538	0.0283	0.0334	0.0386	0.0439	0.0493	0.0548	0.0605	0.0663	0.0722	0.0783
90	2.2994	0.0278	0.0327	0.0378	0.0430	0.0483	0.0538	0.0593	0.0650	0.0708	0.0767
100	2.3452	0.0272	0.0321	0.0371	0.0422	0.0474	0.0527	0.0581	0.0637	0.0694	0.0752
110	2.3912	0.0267	0.0315	0.0364	0.0414	0.0465	0.0517	0.0570	0.0625	0.0681	0.0738
120	2.4366	0.0262	0.0309	0.0357	0.0406	0.0456	0.0507	0.0560	0.0613	0.0668	0.0724
130	2.4820	0.0257	0.0303	0.0350	0.0398	0.0448	0.0498	0.0549	0.0602	0.0656	0.0711
140	2.5272	0.0253	0.0298	0.0344	0.0391	0.0440	0.0489	0.0540	0.0591	0.0644	0.0698
150	2.5727	0.0248	0.0293	0.0338	0.0384	0.0432	0.0480	0.0530	0.0581	0.0633	0.0686
160	2.6171	0.0244	0.0288	0.0332	0.0378	0.0425	0.0472	0.0521	0.0571	0.0622	0.0674
170	2.6624	0.0240	0.0283	0.0327	0.0371	0.0417	0.0464	0.0512	0.0561	0.0611	0.0663
180	2.7071	0.0236	0.0278	0.0321	0.0365	0.0410	0.0457	0.0504	0.0552	0.0601	0.0652
190	2.7518	0.0232	0.0274	0.0316	0.0359	0.0404	0.0449	0.0496	0.0543	0.0592	0.0641
200	2.7954	0.0228	0.0269	0.0311	0.0354	0.0397	0.0442	0.0488	0.0535	0.0582	0.0631

[a]The manufacturer's listing specifies the temperature range for operation.
[b] W/V [agent weight requirements (lb/ft³)] = pounds of agent required per cubic foot of protected volume to produce indicated concentration at temperature specified.

$$W = \frac{V}{s}\left(\frac{C}{100-C}\right)$$

[c]t [temperature (°F)] = design temperature in the hazard area.
[d]s [specific volume (ft³/lb)] = specific volume of HFC-227ea vapor can be approximated by $s = 1.885 + 0.0046t$, where t = temperature (°F).
[e]C [concentration (%)] = volumetric concentration of HFC-227ea in air at the temperature indicated.

Reprinted with permission from NFPA 2001-2018, *Standard on Clean Agent Fire Extinguishing Systems*, Copyright © 2017, National Fire Protection Association, Quincy, MA. This reprinted material is not the complete and official position of the NFPA on the referenced subject, which is represented only by the standard in its entirety which may be obtained through the NFPA website at www.nfpa.org

Halon Systems and Halon Phase Out

Halon, like carbon dioxide, is a fire extinguishing gas that is colorless, odorless, and leaves no residue, but there are several important differences between halon and carbon dioxide. Carbon dioxide extinguishes combustion by physically suffocating the fire: replacing the oxygen necessary for combustion with the carbon dioxide gas. Halon extinguishes combustion by inhibiting the chemical reaction of fuel and oxygen. The actual chemical reaction by which halon extinguishes combustion is not completely known, but this is one rather generic theory that is used in this text to describe the basic difference between the extinguishing action of carbon dioxide and Halon.

Another important difference between Halon and carbon dioxide is the effect of the gas on personnel, and this difference is one that has made halon popular as an extinguishing agent. By replacing sufficient oxygen to suffocate combustion, carbon dioxide also replaces sufficient oxygen to suffocate personnel. Since halon does not depend on oxygen replacement to extinguish a fire, the concentration of halon necessary to extinguish most surface fires is generally not harmful to personnel, an extremely important advantage, especially in total flooding systems. Recall from the section on carbon dioxide systems that it is necessary to have a pre-discharge alarm with a total flooding carbon dioxide system. This was to provide time between the first alarm and discharge for personnel in the area to evacuate because once the discharge commences, the occupants were in danger. With Halon 1301, the most common and popular halon used today, the total flooding discharge concentration for areas such as a computer room or control room is not hazardous to the occupants.

There is a restriction on Halon use, however, that affects its continued popularity. Halon 1301 is a member of a family of halogenated agents that has a detrimental effect on the earth's ozone layer, Halon 1301 production has been frozen by international agreement. At the present time, halon system use is limited, and these systems can now only be used if no other agent can provide protection. Most government agencies and environmentally conscious corporations completely prohibit the installation of new Halon sys-

tems. Accordingly, when a Halon 1301 system is authorized, testing of the total flooding system becomes a major concern.

The testing of total flooding Halon 1301 systems now consists of pressurizing the enclosure to determine the concentration that will be retained for the required time period. This test does not verify that the concentration is within the safe 5–7% range, nor does it confirm that the concentration distribution throughout the enclosure is constant.

Use of Halon 1301

Concentrations of Halon 1301 of between 5% and 7% by volume are levels that can safely be inhaled by personnel for 4–5 min without serious cardiac or central nervous system effects. This means that discharge can commence immediately upon detection with no pre-discharge alarm time delay. Concentrations above 7% can cause slight dizziness or a feeling of being mildly intoxicated. When the concentration passes 10%, mental and physical impairment can occur. Concentrations above 17% are assumed to cause unconsciousness and possible death. When it is necessary to provide protection in an area where personnel cannot cease activity or shut down equipment and evacuate the area immediately upon hearing the fire alarm, Halon 1301 discharge can commence combustion suppression when manually activated. As a result, total flooding Halon 1301 systems have found extensive use in areas like airport control towers and air traffic control centers, where personnel cannot jump up and run for the exit when the fire alarm is activated.

Need for Precise Design

When Halon 1301 is used as a total flooding system in areas where personnel are present, and the concentration must not exceed 7%, it is absolutely necessary that the system be designed and calculated with a high degree of precision. Further, the design of a total flooding Halon 1301 system must be performed by an experienced professional because the flow calculations to provide the proper discharge at each nozzle with all but the simplest systems is extremely complex and sophisticated. These are so complex that they cannot be done manually and must be performed by computer programs (which are checked by national testing laboratories to assure not only their accuracy, but also their limitations).

The purpose of this section is to acquaint the reader with the preliminary specification and design of a system. In no way should a lay person attempt to design any of the gaseous systems described.

Alarm Systems

Halon 1301 total flooding systems are automatically activated by a detection system, and in order to prevent false discharge, the detection system is usually designed to provide all alarms, automatically close self-closing doors to the enclosure by releasing magnetic door holders, and shut down heating, ventilating, and air-conditioning (HVAC) when the first detector is activated. Activation of a second detector releases the discharge.

The detection system provides the alarms, but since there is provision for manual operation of the discharge, it is best to always include in the specifications and cost estimate a pressure switch on the halon discharge piping. This switch will activate all the alarms and control devices that the detection system handles if the detection system is bypassed by manual operation. In lieu of magnetic door holders, self-closing doors may be held open and automatically closed by using pressure trips that are connected by piping to the discharge piping of the system. When the system operates, pressure in the piping activates the pressure trip devices and the door closes. (These pressure trips can also be used with CO_2 systems.) Whether the alarms are bells, horns, or sirens, it is important to have a visual signal, such as a strobe light, to warn personnel who might not hear the alarm or who have a hearing impairment.

An audible and visual alarm are also necessary immediately outside of the protected area to provide individuals outside the room with notification that the halon has discharged. The sign/alarm should post notice not to open doors and to rescue personnel who may need assistance in leaving the area. Responding firefighters also need this notification so they can use self-contained breathing equipment before entering the area.

Rapid Discharge Rate

The total flooding system shall be specified to establish the required concentration in the enclosed area within 10 s, and this concentration is to be maintained for a period of 10 min. One reason for this rapid discharge rate is the fact that Halon 1301 will decompose when exposed to temperatures in the 900 °F range, and the decomposition products not only have an extremely irritating odor but can also be very toxic. By developing the concentration very rapidly, the combustion can usually be suppressed and extinguished before materials exposed to the fire can reach high temperatures. Flaming combustion can be quickly suppressed, thereby reducing the heat of the fire.

The extremely rapid rate of discharge of a total flooding Halon 1301 system presents a condition that must be given careful consideration. When a Halon 1301 total flooding system is used to protect areas where the velocity of the discharge could damage delicate material, a very careful evaluation of the location of the nozzles, the discharge rate, and the type of nozzles must receive top priority.

For example, such a detailed evaluation was not developed for the installation of a Halon 1301 total flooding system in an archives storage room of an art museum. When the discharge test was conducted, the velocity of the discharge blew irreplaceable artifacts off the shelves and damaged them beyond repair. The extremely rapid rate of discharge of a total flooding Halon 1301 system must be given careful consideration.

The velocity of the discharge has been known to blow ceiling tiles off a suspended ceiling support, and the noise of the discharge can be extremely frightening, if unexpected. To counter this, there is customarily one type of alarm when the first detector is activated, and a different type of alarm when the halon is discharged, although the alarm in the protected area is quickly drowned out by the noise from the discharge.

Storage of Halon

Halon 1301 is stored in cylinders as a liquid and, to maintain the Halon 1301 in a liquid state and to provide sufficient pressure to transport the liquid Halon 1301 through the piping to the discharge nozzles. The pressurization is maintained at 360 psi at 70 °F or 600 psi at 70 °F. The 360 psi level is usually specified for systems with relatively short piping designs, and the 600 psi system is primarily used for complex piping designs or where there is considerable distance between the storage cylinders and the hazard.

If it is not possible to replenish the supply of Halon 1301 cylinders within a reasonable time (usually 24 h) or if the hazard is extremely valuable to the operation of the facility, or the hazard has the potential for recurring fires, it is necessary to furnish reserve supply cylinders permanently piped into the system. The operation of a switch will change the system supply from the empty primary cylinders to the reserve cylinders. A check valve in each supply from the primary and reserve cylinders to the system piping will prevent the charge cylinders from discharging into the empty cylinders.

Control Switches for Storage

Manual pull stations are provided to trigger the system manually, and activation of a pull station transmits a signal to the cylinder discharge devices. Activation of a manual pull station will cause immediate discharge of the halon, and at the same time set off all alarms and control devices.

Each system has a control panel that receives signals from the detection system and transmits signals to the cylinder discharge devices, alarm system, and control systems.

Abort switches may be requested by an owner to abort, or put on hold, the discharge of the halon after a detector has activated the alarms. Automatic time delays to delay the discharge of the halon following activation of the detection system are acceptable, but the use of a device such as an abort switch to prevent halon discharge until a fire situation is confirmed is not acceptable, and usually not permitted by the insurance carrier.

When an abort switch is installed, it should be a dead man switch. A dead man switch must be held in the operating position by an individual, and once released it will automatically activate the halon discharge. If an individual hears the alarm when the first detector is activated and does not see any smoke or fire condition, he or she can hit the abort switch and prevent the halon discharge.

A note of caution If some individual hits the abort switch, the only way they can release the switch without discharging the halon is to reset the panel that controls the system. If they are alone in the area, and it is obvious that a non-fire condition activated the detector, the abort switch should be within reach of a telephone or the control panel.

Never use a positive type abort switch similar to a toggle switch. There is too much probability that the switch would be left in the off position rendering the system inoperative.

Placement of Halon Cylinders

Halon 1301 cylinders should be stored outside the hazard area and secured with brackets designed to hold the cylinders. The storage cylinders should be safe from mechanical and physical damage, and placed in an area where the temperature will not exceed 130 °F nor fall below −20 F.

Storage Conditions

Systems with the halon cylinders located outside the hazard area, with the extinguishing agent piped to the discharge nozzles, are referred to as engineered systems.

In other Halon 1301 systems, the individual cylinders are installed in the hazard area. These containers may be either cylindrical or spherical in shape, and can be installed on the wall, above the suspended ceiling, or below the raised floor. Each cylinder is a self-contained system with supply, cylinder actuator, and discharge nozzle. The requirement for piping from the cylinder to the discharge nozzle may consist of a short nipple, or the nozzle may be part of the cylinder itself with no piping involved. Systems with the supply cylinders located in the hazard area are commonly referred to as modular systems.

Estimating Approximate Amounts

During the design of an engineered Halon 1301 system, the questions most frequently asked by the architect or the mechanical engineer are: How many Halon 1301 cylinders? How big are the cylinders (to allocate storage space)?

Below is a simple method by which to estimate the approximate amount of Halon 1301 required to protect the hazard with a 5–7% concentration:

1. Calculate the cubic footage of the enclosure. If there is a raised floor, as found in the majority of computer rooms, use the height of the area from the bottom of the suspended ceiling to the floor below the raised floor. The area above the suspended ceiling does not have to be included in the calculation for total cubic footage unless there are combustible materials above the suspended ceiling.
2. Once the cubic footage has been determined, multiply this figure by 0.025, and the answer will be the estimated amount of Halon 1301 in pounds.
3. Consult a halon manufacturers catalog for the cylinder capacities in pounds, and once the appropriate cylinder is selected, the size will be indicated.

If the weight of Halon 1301 is considerable, it will be necessary to divide up the total pounds required among several cylinders, if reserve supply is required an equal number of equal size cylinders will occupy the storage space.

This formula for estimating the required number of pounds of Halon 1301 includes about a 20% safety factor to account for normally encountered loss of halon concentration through various unsealable openings. This type of unsealable opening occurs below doors: exhaust fans that are shut down when the detection system is activated may continue to rotate, or ducts that have dosed dampers during discharge of the halon often have a certain amount of leakage.

Loss of gas through these small openings is accented by the tremendous rate of discharge, so this 20% safety factor is necessary in figuring the approximate number of pounds of Halon 1301 required. Large openings that cannot be closed prior to the discharge must be carefully considered in the design calculations.

In areas with underfloor spaces, do not attempt to protect the underfloor space with above floor system. The underfloor space in usually used as ventilation for the machines on the raised floor, and a considerable amount of halon discharged in the area above the floor will spill into the underfloor space, but some of this halon will be sucked up into the machines and the below floor concentration will be diluted. If the underfloor space is to be protected, it is best to install both detection and discharge nozzles at the ceiling for the above floor area, and detection and discharge nozzles below the raised floor. Activation of the detection either above the floor at the ceiling or below the raised floor must activate the entire system. The discharge nozzles at the ceiling for the area above the raised floor, and discharge nozzles below the raised floor, must activate simultaneously.

In requesting bids or detailed system installation design and calculations, do not neglect any openings in the area that cannot be dampered, shut down, or dosed. This includes open drains and cable shifts. Also, be absolutely certain that raised floor spaces do not extend beyond the above floor partitions, as loss of halon into adjacent unprotected areas or areas beyond the scope of the system will dilute the concentration below that required to extinguish the combustion.

Partial Total Flooding Halon 1301 Systems

Halon 1301 is an expensive agent, and when the equipment to be protected is in a room of considerable height, for example 40′, and the equipment height is only 15′, the total flood of the room can be calculated to cover the equipment plus 5′–10′ (above). Since Halon 1301 is considerably heavier than air, it will settle to the lower areas of a room. To ensure the accuracy of the calculations, design, and placement of the nozzles, experienced Halon 1301 design professionals must engineer the system.

An example of partial total flooding is the protection of aircraft flight simulators. The area containing these simulators is considerably higher than the simulators. The Halon 1301 is discharged to provide a 6% concentration to protect the sophisticated electronic equipment associated with the simulators, to a height of approximately 10′ (above the simulators).

Dry Chemical Systems

Dry chemical is a generic term because it includes several different types of dry chemical powders in extremely small particles. Dry chemicals have been used in hand extinguishers since the 1920s, but their use in fixed pipe extinguishing systems is relatively recent. The original and still most well known dry chemical is sodium bicarbonate, but there are other dry chemicals that have gained tremendous public acceptance. Potassium bicarbonate, also known as Purple K, is used extensively on hydrocarbon fires because it is almost three times as effective as sodium bicarbonate. Monoammonium phosphate is an all purpose agent, and is used in the majority of home fire extinguishers.

Classification of Dry Chemicals

Before proceeding with a discussion of dry chemicals and dry chemical systems, it is necessary to reiterate the lettering system used to identify different classes of combustibles:

Class A combustibles are ordinary combustibles such as wood, paper, cloth, rubber, and some plastics.
Class B combustibles are flammable liquids, paints, grease, solvents, etc.
Class C is not combustible, but indicates that the agent is electrically nonconductive, and can therefore be safely discharged on energized electrical equipment, devices, and wiring.
Class D refers to combustible metals such as magnesium, sodium, and potassium. Agents

used for extinguishment of Class D combustibles are not referred to as a dry chemicals, but as dry powders.

Sodium bicarbonate is basically a Class B extinguishing agent, but it can handle small Class A fires that are on the surface of the combustible. Purple K is primarily a Class B extinguishing agent because of its efficiency in fighting flammable liquid fuel spill fires. Monoammonium phosphate is a multi-purpose agent because it is an agent used to extinguish Class A and Class B combustibles, and is safe when used on energized electrical equipment, so it is an A/B/C agent. Dry chemical systems can be either total flooding or local application type.

The design criteria for a dry chemical total enflooding system closely follows the requirements for a carbon dioxide total flood system: (1) discharge of the agent into an enclosed space, (2) system activation automatically by a detection system with manual activation potential, (3) doors, windows, and dampers automatically closed, and (4) HVAC (heating, ventilating, air-conditioning) system shut down when system is activated.

Here the similarity ceases with a dry chemical total flooding system. With gaseous systems, the required concentration in the enclosure must be maintained for a specified time period to prevent re-ignition of the combustion. Dry chemical concentration must be obtained in all sections of the enclosure within 30 s, but there are no requirements for maintaining the concentration because the dry chemical settles quickly. For this reason, when designing a dry chemical system sources of reignition must be evaluated along with the hazard itself and eliminated if at all possible.

The total flooding concentration for various hazards is mandated for gaseous systems, but this is not possible with dry chemicals. The quantity of dry chemical, the rate of flow in the system, and the piping configuration for nozzle positioning are all dependent on the specific dry chemical used and the hazard being protected. This text will provide the basic criteria for a total flooding dry chemical system, but experienced professionals must be consulted once the preliminary plans have been established.

Safety to Personnel

Safety to personnel is a high priority concern with dry chemical total flooding systems. The discharge of dry chemical systems can cause loss of visibility, and produce difficulty in breathing, resulting in a panic condition. If personnel are present in the area protected with a dry chemical total flooding system, a pre-discharge alarm should be provided to allow evacuation before discharge.

Dry Chemical System Design

Most fixed pipe dry chemical total flooding systems, called engineered systems, are two cylinder systems: one cylinder contains the dry chemical, and the other contains either nitrogen or carbon dioxide. When the detection system is activated, or the system is tripped manually, the nitrogen or carbon dioxide is automatically discharged into the dry chemical cylinder, pressurizing the dry chemical which is then carried through the piping to the discharge nozzles. However, this is not quite as simple an operation as it sounds. Before the dry chemical can be transported through the piping in a free flowing manner, the gas/dry chemical mixture in the cylinder must be in a fluid state. In this condition, the mixture is said to be fluidized. The dry chemical must not be fluidized before it can be released to the piping, but also the pressure in the dry chemical cylinder must build up to the design pressure required to transport the fluidized mixture through the piping, arriving at the nozzles at the required pressure.

To equalize the pressure in the dry chemical cylinder, a pressure regulator is usually provided to control the pressure of the nitrogen or carbon dioxide into the dry chemical cylinder. To accomplish the equalization of the pressure in the dry chemical cylinder, and to assure the fluidization of the dry chemical in the cylinder, the dry chemical cylinder has a bursting disc which seals it off from the piping until the calculated pressure is reached. When the predetermined pressure is attained in the cylinder, and the dry chemical has been fluidized, the disc will rupture and the dry chemical/gas mixture will enter the piping and flow to the discharge nozzles. The bursting disc

(between the empty piping and the dry chemical cylinder) also prevents entry of moist air, condensation, and caking of the dry chemical.

The fluidized mixture is transported through the empty piping at extremely high velocities, upwards of 125 feet per second, at pressures of up to 350 psi. This high velocity and pressure, combined with the fact that although the dry chemical is technically a powder (in the gas/agent mixture) when in a solid state, produces a tremendous thrust and force at elbows and tees. The dry chemical system must be designed and installed with the necessary strength to contain this force. The fact that the dry chemical is a solid makes it rather easy to visualize how the dry chemical powder by centrifugal force separates from the gas at an elbow and follows the curve. With the dry chemical and gas mixture flowing through the piping at tremendous velocity, and after being thrown against the outside curvature of the elbow (if the next fitting is a tee), the dry chemical may rush past the tee outlet hugging the outside of the straight part of the tee. To give the dry chemical a chance to re-mix with the gas before it reaches the tee, a straight piece of piping, not less than 20 times the pipe diameter in length, must be installed between the elbow and the tee.

If space limitations will not allow this length of pipe, a device called a venturi assembly is installed between the elbow and the tee. This venturi is basically a restricted throat which allows the dry chemical to re-mix before entering the tee.

With tees on a straight run of pipe, the dry chemical traveling at a high velocity in the straight pipe can pass over the outlet of the tee. To overcome this effect, tees are most effective when they are installed bullheaded -discharging into the side of the outlet of the tee.

Tests

The optimum amount of dry chemical and the rate of discharge for specific hazard have been determined by experiment and test for total flooding systems. The quantity of dry chemical, rate of flow, discharge rate, and nozzle placement resulting from these experimental tests are plotted on graphs.

The tests that provide information for these graphs are based on using different flammable liquid surface fire areas, the amount of dry chemical required to extinguish the combustion, and the minimum rate of discharge necessary to accomplish the extinguishment. Each manufacturer must conduct tests since every piece of equipment-nozzles, devices, and dry chemical compounds-affect the flow rate discharge capability for their specific system. The graphs developed from these tests can be obtained from the various dry chemical manufacturers for use in specifying and preliminary design of dry chemical total flooding systems. The following formula was developed from the tests conducted by one manufacturer and is offered here to provide the generic concept of testing for total flooding dry chemical systems. It is imperative that this formula be considered as an example only, and should not be used for all situations, all types of dry chemicals, or all manufacturers systems: The weight of dry chemical in pounds for a total flooding system can be determined by multiplying the cubic feet of the enclosure by 0.0385. As with all dry chemical total flooding systems, this is the gross cubic feet of the enclosure less the volume of permanent equipment or other permanent fixtures that occupy a considerable volume in the enclosure: This formula is presented for example only and should not be used for any preliminary calculations that will become part of a specification or cost estimate. Each specific total flooding dry chemical system calculation must be based on the manufacturers formulas and graphs for the particular hazard and enclosure.

Local Application Dry Chemical Systems

Local application systems are used when there is no enclosure around the hazard or when the hazard is located in an area enclosure where total flooding would be impractical. As with other local application systems, the dry chemical discharges directly onto the hazard. Dry chemical local application systems are used extensively to protect solvent and paint dip tanks, open tanks of flammable liquids, gasoline, and flammable liquid racks. Dry chemical local application systems have also been used to protect oil insulated electric transformers located in transformer vaults and exteriors transformers. When the system protects a hazard located outside, additional dry chemicals must be included to compensate for loss due to wind currents carrying some of the discharge away from the hazard. As with total flooding systems, the design of local application systems depends on the type of dry chemical, the rate that the dry chemical is discharged, the nozzle type used, and the velocity of the dry chemical. All of these design and calculation factors, including the amount of dry chemical required for the hazard, are determined by experimental testing by manufacturers and testing laboratories.

Local application systems are usually activated automatically by a detection system with the provision for manual operation of the system, and the majority of large systems are supplied from the dry chemical cylinder pressurized from the nitrogen or carbon dioxide cylinder. The design of the piping for a local application system is similar to that described for the dry chemical total flooding system. Pipe lengths, pipe sizing, number and size of fittings, type, and placement of nozzles are critical. The local application system will experience the same, if not greater, velocities and pressures, requiring application of substantial pipe supports and installations to resist the extremely high thrust conditions.

Dry Chemical Hand Hose Systems

Where large fuel spill fires are a potential hazard, dry chemical hand hose lines perform as exceptional firefighting equipment. Hand hose lines are supplied from a dry chemical storage cylinder with a separate nitrogen or carbon dioxide cylinder to pressurize the dry chemical. Pressure forces it through the hose to the discharge nozzle, and the hose operator has the capability of controlling the discharge with an on-off nozzle valve.

Each hose station must have a supply of dry chemicals capable of furnishing a discharge for a minimum of 30 s. The hose stations can be individual, self-contained units with the dry chemical supply dedicated to the individual hose station, or the hose supply can be furnished from the same dry chemical supply that is furnished for a local application or total flooding system. The supply for both the system and the hose stations is designed using selector valves that direct the discharge through the piping. The selector valve activation determines which piping system will receive the dry chemical discharge.

The flow rate of the dry chemical is a function of the hose size, hose length, and nozzle. The average flow rate is approximately eight pounds per second for a 1″ hose, 15 pounds per second through a 1/2″ hose, and three pounds per second through a 3/4″ hose. For example, with 50′ of hose and a nozzle, the flow rate is approximately 2.7 pounds per second, with a range of about 25, with a 150 pound supply of dry chemicals, which will discharge for approximately 49 s.

The strong cylinders for supplying hose systems, and total flooding or local application systems plus hose stations, can range from 125 to 3000 pounds of dry chemical.

Pre-Engineered Dry Chemical Systems

The discussion of dry chemical system has presented the many and varied difficulties in designing and calculating engineered total flooding and local application dry chemical systems. Pre-engineered dry chemical systems have already been engineered by professionals experienced in the design and calculation of dry chemical systems. To protect relatively small hazards that have similar construction configurations, dry chemical systems are tested by the manufacturer using actual fire tests for various hazards. These tests and the test results are confirmed and verified by Underwriters Laboratories, a nationally recognized testing laboratory.

The design of the system including the number and type of nozzles, quantity of any chemical, type of dry chemical, method of operation, maximum and minimum pipe lengths, number and type of fittings, rate of application, and minimum effective time of discharge, are all part and parcel of the UL listing given by the Underwriters Laboratories. A pre-engineered system must be installed within the parameters of the UL listing which are covered in the manufacturers installa-

tion manual. This manual is an integral part of the UL listing, and the system components listed in the manufacturers manual can be used.

Although pre-engineered dry chemical systems may be total flooding or local application type, the most common pre-engineered systems are used for protecting kitchen hoods and hood exhaust ducts. Kitchen hood systems are automatically activated by a conventional detection system or by fusible link (similar to a sprinkler head fusible link) which releases the actuating device when heat fuses the element and the link separates. There is always a manual means of discharging the system, and on some small systems manual activation is the only means of system activation. Kitchen hood pre-engineered dry chemical systems must be equipped with an automatic device to shut off fuel and power to the cooking appliances when the system is discharged.

Pre-engineered dry chemical systems are very effective in protecting laboratory hoods, paint and solvent dip tanks, and the drain boards for dip tanks, to name a few hazards. They are also an economical means of protection for hazards that can be protected with one of these UL listed systems. The sophisticated design, calculations, and actual fire testing has already been performed, and if the guidelines, mandates, equipment, devices, and design included in the UL listing are used and followed, hazard protection with dry chemicals is readily available.

References

EPA. "Carbon Dioxide as a Fire Suppressant: Examining the Risks." EPA. <https://www.epa.gov/sites/production/files/2015-06/documents/CO2report.pdf)>. 2015.

EPA, Emissions. "Understanding Global Warming Potentials." <https://www.epa.gov/ghgemissions/understanding-global-warming-potentials>. 2017.

EPA, OLP. "Basic Ozone Layer Science." <https://www.epa.gov/ozone-layer-protection/basic-ozone-layer-science>. 2017.

Hurley, Morgan J., Gottuk, Daniel T. et al. *SFPE Handbook of Fire Protection Engineering*. Springer, New York. 2015.

Stat-X. "Stat-X Products." <https://www.statx.com/products/>. 2005.

Wickham, Robert T. "Review of the Use of Carbon Dioxide Total Flooding Fire Extinguishing Systems." 2003.

Regulatory Agencies, Authorities and Organizations

13

Regulatory Agencies and Authorities

Understanding regulatory agencies and the authorities having jurisdiction is key to designing, engineering, specifying, estimating, and installing fire protection systems.

A conversational, if not working, knowledge of fire protection systems and the equipment that comprises those systems is important to all construction and facilities management disciplines. However, this chapter covers a subject that separates the professional and experienced fire protection engineer from the other disciplines involved in fire protection projects.

The guides provided by NFPA, Factory Mutual, and other organizations furnish the specifics of system design, but almost every one of these guides makes the statement in so many words, subject to the approval of the authority having jurisdiction. Who are the authorities having jurisdiction? On some projects they are clearly identified, but on many others this phrase used in a specification indicates the specification writer's uncertainty, in which case the project could be headed for trouble.

The authority having jurisdiction may be the insurance carrier, or the specifications may require that the system be installed in accordance with state and local codes. These codes may reference ordinances and agencies that can seriously impact a project. Designers should be forewarned

that even a very tight specification and reference to compliance with NFPA and/or FM requirements can be destroyed by making blanket overall statements like those mentioned above.

Today's economy-mindedness has inevitably reduced the quality control that existed in the past. It was once common practice to submit preliminary fire protection plans and design criteria to the authorities, as well as to the insurance interests having jurisdiction, for their review, comments, and approval. Only after these plans were returned with the comments and stamp of approval did the final design begin. Today this process of coordination may be at an opposite extreme, with design and construction occurring simultaneously.

National Fire Protection Association

The National Fire Protection Association is not of itself an authority having jurisdiction, because the NFPA does not approve design plans or conduct tests or inspections of system installations. The NFPA does not approve or certify equipment or materials, nor does the NFPA single out or approve testing laboratories that evaluate equipment, materials, or systems.

The authority, or the insurance organization that does have jurisdiction to approve plans and installations may, however, base approval of the plans and the installation on compliance with NFPA Fire Code requirements. As such, the

© Springer International Publishing AG, part of Springer Nature 2019
R. C. Till, J. W. Coon, *Fire Protection*, https://doi.org/10.1007/978-3-319-90844-1_13

NFPA is the authority, but only within the framework of the local code (in 2018 usually a variant of the International Building Code (IBC).

Understanding the origin of the NFPA Fire Codes is important because it explains the reasons and significance of codes in fire protection today. When automatic sprinklers were first developed beyond the experimental stage, and began to be installed in buildings, several pipe sizing methods were in use. Differences in rules produced installations of all sizes and configurations. There was no proven way to ascertain the effectiveness or reliability of these installations, since there was no clear track record.

Insurance companies were providing premium savings for sprinklered buildings, yet with no established standard of installation, they could not establish a foundation of sprinkler installation reliability. As a result, a group of representatives from insurance carriers met in Boston in 1865 to establish a uniform standard for sprinkler installations. Their purpose was to obtain the installation uniformity, system performance reliability, and efficiency needed to establish realistic sprinkler premium rates. This meeting resulted in the first standard based on the consensus of opinions of experienced personnel for the design and installation of automatic sprinkler systems. This first consensus led directly to the birth of the NFPA in 1896.

The small group that established the first technical committee has grown into over 9000 volunteer committee members, offering a vast source of experience required to develop, monitor, and periodically update over 300 codes and standards. These NFPA technical committees have jurisdiction over basic NFPA documents: the code, the Standard, the Recommended Practice, the Guide, and the Handbook.

A *code* is a model, a set of rules that knowledgeable people recommend for others to follow. It is not a law but can be adopted into law.

A *standard* tends be a more detailed elaboration, the nuts and bolts of meeting a code.

NFPA Codes and Standards mandate requirements with the use of the word shall. These requirements are, in most cases, detailed and specific design and installation requirements. NFPAs Recommended Practice and Guide are exactly what the names imply, and the word *should* is used in these documents in lieu of the word *shall*. Handbooks elaborate on the document that they refer to.

A note of caution: The majority of the NFPA Standards have a section at the back of the Standard called the Annex: The Annex is offered as part of the Standard for information only and is not part of the body of the Standard. The words *should* and *may* appear in the Annex. Since Annex items are not part of the body of the Standard, they are not mandatory requirements, but are for information only, and are used to explain a section of the Standard.

According to the administrative authority of NFPA Technical Committees and the Fire Codes, all Standards, Recommended Practice, or Guides must be reviewed, updated, or reaffirmed in their present content by the Technical Committee in charge of the document every five years. In the past a vertical line beside a Section or paragraph or sentence in a Fire Code, Standard, Recommended Practice, or Guide is an indication that this is new wording since the publication of the last document. New ways of denoting new content will be appearing soon.

To this day, even though electronic versions are readily available, paper versions of the codes and standards are issued as the National Fire Codes Archive Set. One reason is that the professional engineering exams still require the use of paper codes in the testing area. The numerous hard copy volumes of Fire Codes contain all the current Standards, Recommended Practices, and Guides, but the date of the current issue of the Fire Code volume is not necessarily the latest issue of the particular document being used. For example, the 1989 set of National Fire Codes, with 1989 printed on each volume, may contain Standards dated 1988 or 1987, etc.

If the Standard, Recommended Practice, or Guide is being referenced by date in the specification, be sure to check the date of the specific document. This may sound like an unnecessary caution, but changes in an updated document, particularly a Code or Standard, can cause serious problems if the date used in the specification does not coincide with the actual date of the document.

The NFPA Code requirements effectively become the jurisdictional authority when they are used as a basis of acceptance or approval of a system design or installation by the proper authority or the insurance interest having jurisdiction. NFPA Codes actually become law when referenced by number and date of issue in a building code and can be considered as a law when referenced in a specification, by a fire marshal, government agency, local fire prevention bureau, fire department, health department, or any other similar authority or insurance carrier having jurisdiction. The NFPA may reference other Standards that had some bearing on a particular Standard under consideration. NFPA references these other Standards in the body of their Standard. In such cases, it becomes mandatory to use referenced Standards in the design and installation of the system, if applicable to the design and installation of the particular system.

National Electric Code

The National Electric Code (NEC) warrants individual recognition because this NFPA Fire Code is considered to be law throughout the United States. The NEC has mandatory requirements that affect all types of electrical installations, of which fire detection and alarm systems are only one facet of the diversified code requirements. One section of the NEC contains the design criteria and material, equipment, and devices to be used in areas subject to potential flammable liquid vapor or dust explosions. There is also a section on the location of explosion-proof devices and equipment relative to a fueled aircraft in an aircraft hangar. The important feature of the NEC is that it must be considered in the evaluation of every detection and suppression system and every hazard condition or situation.

FM Global

The Factory Mutuals were founded in 1835 based on the conclusion that large industrial and commercial companies needed to provide mutual fire insurance coverage and loss prevention. In the past the Factory Mutual System consisted of three insurance firms-Allendale Insurance, Arkwright Insurance Company, and Protection Mutual Insurance Company. As of 1999 the Factory Mutual System is now referred to as FM Global.

The Factory Mutual (FM) Research Corporation conducts research, and out of this research department have come equipment, devices, and systems that have benefited the entire fire protection industry and profession. One function of the FM Research Corporation is the testing of equipment, devices, and systems to determine if their reliability and efficiency would warrant the FM Approval label. The indication that a device, piece of equipment, or system is FM Approved provides the user with the confidence that it has been thoroughly tested and found to be worthy of use in a fire protection design.

The Factory Mutual Approval Guide is a publication that is constantly updated, and lists all the products, devices, equipment, and systems that are FM approved. The Approval Guide not only lists FM approved items, but also details the material and installation criteria. In some instances the detailed specifications outline the parameters of usage to remain within the FM Approval.

The Factory Mutual research activities are located on the East coast (Norwood, Massachusetts), and the facilities are staffed by experts in many fields. Factory Mutual Research makes these facilities and laboratories available to private companies and organizations on a contract basis.

Since FM policies can cover a multitude of loss coverages, their testing and requirements go far beyond fire protection. FM requirements include windstorm and hail, explosions, vandalism, sprinkler leakage, lightning, and loss or damage from furnace malfunction. FM publishes a great deal of literature on requirements for protecting and preventing losses from these hazards.

Factory Mutual has always limited their insurance coverage to highly protected risks, and their requirements reflect this environment. A very important member of the FM family is the Factory Mutual Engineering Department. These departments are staffed by experienced professional fire protection engineers who review the design and installation plans on all FM projects.

NFPA has their Fire Codes, and Factory Mutual publishes their requirements in the Factory Mutual Data Sheets. Make no mistake, if the project is a Factory Mutual client, use the Factory Mutual Data Sheets to obtain the design requirements. Some Factory Mutual Data Sheets will use an NFPA Standard word-for-word, but FM will always add in bold print any additions to, or changes from, the NFPA requirements.

In addition, FM has a staff of experienced, professional, field personnel who conduct the testing of new systems; these field personnel use the testing requirements of Factory Mutual. When specifying the testing of the systems, be sure that the specification section is based on FM requirements. This aspect of a project also can have a cost impact, because the installation contractor may be required to furnish additional labor, material, and equipment to satisfy the test requirements.

Architects and engineers who have a client that is, or will be, insured by Factory Mutual should always arrange meetings with Factory Mutual personnel at the start of the project. Factory Mutual offers the FM Data Sheets to indicate their requirements, but every FM project protection system is tailor-made for the specific hazard and/or facility. FM may require other design features not covered in the Data Sheets. FM fire insurance premiums are generally very low because they require the most effective and reliable protection that is available to protect their risk.

In summary, do not try to outguess the protection requirements on an FM insured risk, and do not look at the Data Sheets as the last word in requirements for the design. The only sure way to provide the client with a Factory Mutual approved system is an exchange of communication with FM personnel.

These meetings with FM are not being recommended just for fire protection discussions. On new building construction design, they must cover all aspects of loss prevention construction, and architects and engineers of all disciplines on the design team should participate.

If FM is brought in late in the design process, the architect and engineer may find that they have overdesigned or underdesigned the facility, including the fire protection.

Building Codes

Building Codes are the product of a jurisdictional authority having the power of the law. This power is administered by the building code officials of a Building Code Department.

Municipal building codes, and those codes that have been adopted to cover an entire state, provide the requirements to establish an occupancy classification, and many of the construction features that apply to the various occupancy classifications.

The major problem for the architect, engineer, and fire protection engineer is that there are many sources of building codes, and each code has its own set of requirements and mandates. Three of the many building codes that were most frequently used before they were consolidated into the International Building Code (IBC) by the International Code Council (ICC) are: The Uniform Building Code (UBC), produced by the International Conference of Building Officials (ICBO), the BOCA Building Code (BOCA), produced by the Building Officials and Code Administrators (BOCA), and the Southern Building Code (SBC), produced by the Southern Building Code Congress.

These model codes were finally consolidated into the IBC in the year 2000, when it became clear that three sets of building codes were becoming untenable to support, and the discrepancies between the codes were costing builders money. After 2000 the other codes were no longer supported. A copy of the Code, modified for New York City, is available online at https://www1.nyc.gov/site/buildings/codes/2014-construction-codes.page.

The NFPA has also put forward a building code, NFPA 5000, however it has not had a high adoption rate in the US. Access is available for free at the NFPA web site (http://www.nfpa.org).

The important thing to remember in dealing with building codes is the fact that they are written to cover a multitude of different types of facilities and construction types. Every project is different, with different occupancies, and different types and numbers of occupants. At times it becomes extremely difficult to fix an occupancy classification for a particular facility, and interpolating the various requirements of the code to provide compliance with the code requirements for the specific facility may seem like attempting to fit a square peg into a round hole.

It has been said that 80% of building codes relate to property protection and life safety, and therefore it is important to evaluate the requirements of the code having jurisdiction prior to a determination of how much fire protection and what type is required. In the majority of building codes, the occupancy classification will determine whether or not fire protection is required, or what type of protection is required.

An important point in dealing with building codes is best explained by a simple example:

> In evaluating the building code having jurisdiction, it is determined that based on the occupancy classification and building configuration, smoke and heat venting at the roof elevation is required. In the section of the code pertaining to smoke and heat venting, specific requirements are presented regarding spacing of the vents and the number of square feet of venting area required. In this case, the building owner attempted to justify the use of power venting to substitute for some of the mechanical vents. The owner, in effect, asked the Board for permission to provide an equivalent amount of venting by alternate means.

The fact is, the designer is violating the code and the law if the decision is made to reduce the number or increase the spacing of vents, if this will result in a reduction of the venting area.

There have been instances where the authority to modify a code requirement consisted of the oral approval of a building inspector, or even a building official. By accepting this authority, the architect and/or engineer is not only violating the law, but opening the door to liability problems.

If there is some requirement in the code that presents a tremendous hardship from a design, operational, or cost standpoint, and the architect/engineer feels that there is ample justification to seek a variance to the code requirement, the only recourse is to apply to the building code department for a variance. Application for a variance is really an appeal to appear before a Board of Appeals, present the justifications, and convince the Board that the variance will not lessen life safety or property protection. This justification may take the form of alternate methods of providing an equal measure of life safety and property protection-alternate methods that may not comply with the letter of the code, but do satisfy the intent.

The Board is made up of architects, engineers, and contractors who are not employed by the Building Code Department. The Fire Chief and/or a Building Code Official will be able to present their recommendations to the Board, but will not have a voting privilege.

It is important to keep in mind that a Board of Appeals does not have the power to change or delete a code requirement, any more than a review board of this type has the power to change or delete a law, regardless of the circumstances. The Board of Appeals only determines the suitability of the alternate methods of construction or materials or provides a reasonable interpretation of a code requirement.

The Building Official has the authorization to enforce all the provisions of the code, with the power of a law enforcement officer. The Building Official can and does exercise this power by refusing to grant an Occupancy Permit if the facility does not meet the code requirements. This is why it is so important to design and construct a facility that meets all the requirements of the code, because there is hardly any other cost impact that affects a client more severely than to have his facility completed and not be able to obtain an occupancy permit after the building inspector finds a serious code violation, or a deviation from the plans approved by the Building Code Department.

The architect, engineer, and/or client must submit plans, specifications, and computations to the Building Department in order to obtain a permit. This permit may be for construction of a part of the structure, or it may be a permit for all the construction features. The cost of the permit is usually based on the overall project cost.

From a fire protection standpoint, the occupancy classification may or may not determine the fire protection required. Some building configurations mandate a requirement for a suppression system regardless of the occupancy classification. It becomes necessary to carefully evaluate the code requirements relative to the architectural design of the structure. Understanding occupancy classification is important. Take the following example:

> A building classified as an office occupancy, and having the size, height, and building materials under this classification did not require the installation of a sprinkler system. Once the project was well-developed and the budgets established, it was discovered that one section of this same building code stated that all occupancy classifications with a basement or story that exceeded 1500 sq. ft that did not have window openings of a certain size, and spacing, located in walls that were entirely above the ground level, were required to be protected with automatic sprinklers.
>
> The facility in the example had a basement that exceeded the 1500 sq. ft, and being below grade, did not have any windows. It was, therefore, necessary to install a sprinkler system. Windows are required for access by the fire department to facilitate firefighting and rescue from the exterior of the building; therefore, where windows are not possible (below grade), a sprinkler system and detection/alarm may be deemed necessary.

It is important to determine the date of the code being applied. This is especially true when writing a specification. Some building code departments may not use the current issue, and also if the project has a lengthy construction period, a new code issue may have required changes or additions.

Most building codes have what are commonly termed "trade-offs." These trade-offs involve the allowance of construction features if a sprinkler system is installed. In other words, if an occupancy classification and type of construction material limits the allowed square footage or height, either may be increased if a sprinkler system is added to the building design. Some of these trade-offs can be very cost-effective, but

they should be studied carefully because most building codes state that the sprinkler system cannot be used to accomplish a construction trade off if such a system is already required for the project.

Most building codes provide Fire Codes that have the same legal authority as the building code used by the municipality. The Fire Chief or Fire Prevention Bureau of the municipality usually assumes jurisdiction over the Fire Code. Fire Codes in general cover special hazards such as combustible and flammable liquids, combustible fibers spray finishing, explosives, and gases, but they also encompass fire prevention and a multitude of processes that go beyond the building code requirements.

The fire department may have the authority to issue permits for especially hazardous processes or storage of hazardous materials and may have the power of law when fire inspectors perform inspections of these facilities.

It is always extremely important to conduct a careful evaluation of the Fire Code as well as the Building Code. In most cases the Fire Code will follow the requirements of the Building Code, but there are many requirements that may only be found in the Fire Code, and the architect, engineer, and contractor must, by law, comply with both the Building Code and Fire Code when both have been adopted by the municipality.

Types of Buildings: General

The codes have specific names for types of buildings and construction types. However, we can just think of them mainly as residential and commercial (office and industrial). There are other specialty buildings such as churches, hospitals, prisons, and sports arenas, all with their own special properties, and similarities to these structures.

New Versus Existing Buildings

After most fire incidents, it may be stated that buildings were not "up to code". The fact is that unless there are major modifications made to a building, it is only required to be "up to the code" adopted in the year it was constructed. Buildings are not reviewed for current code compliance on any type of timely basis. This is a very important fact.

Buildings are generally only required to be updated to current codes if there are radical changes made to the structure. Smaller changes can still result in the introduction of plastics and other bad actors that may not have been considered when the original building was constructed.

Fire Marshal and Fire Prevention Bureau

Discussions of the Fire Marshal and the Fire Prevention Bureau as the authority having jurisdiction must be integrated, because in many cases this authority is one and the same. This section follows the dissertation on the Fire Code because the Fire Code is a basic regulatory tool of the Fire Marshal and the Fire Prevention Bureau.

Not every state has a Fire Marshal, but all do have an authority with the same responsibilities and power of law regarding fire protection and prevention, whether or not this individual carries the title of Fire Marshal.

In some states the office of the State Fire Marshal is a separate department of state government, while in others, the State Fire Marshal's office is under the supervision of the Department of Insurance. In another example, the State Fire Marshal's office is a separate government department reporting directly to the Governor. Regardless, it is important to understand that this authority in every state demands the attention of the architect, engineer, and contractor if projects within the parameters of state jurisdiction for fire safety are to be approved by all authorities having jurisdiction.

The duties and authority of a fire marshal are, in general, to enforce the law regarding storage, sale and use of combustibles and explosives; the means and adequacy of egress for institutions and places of assembly; the investigation of fires as to their causes, and; the suppression and investigation of arson. In addition, most fire marshals are responsible for performing fire prevention inspections.

These duties vary from state to state, but it is safe to say that if a project involves a public building, and especially if the public building is a place of assembly for large numbers of people, or an institutional facility, such as a hospital or nursing home, or for the use or storage of flammable, combustible liquids, or explosives, the Fire Marshal's office will be involved. The project approval must include the Fire Marshal's office or the agency acting under the jurisdiction of the Fire Marshal.

These state agencies usually have jurisdictional parameters beyond that of municipal or county organizations, but the authority to perform as agents of the state on a local, municipal, or county level can be granted by the state. This delegated authority covers specified areas such as inspection, investigations, and enforcement.

On this local level, the municipality or county can delegate authority, by law, to a Fire Chief, or in some instances, to the head of the fire department or fire district. This individual is usually given the title of Fire Marshal as well as the authority to delegate the responsibility of carrying out the mandates regarding fire prevention and protection to an individual (customarily a Fire Chief) or, depending on the size of the fire department, to a division within the fire department. Many states with large municipalities or highly organized counties or fire districts within the state have a state fire marshal or agency with similar authority, and each large municipality, county, or fire district has a local fire marshal and/or a Fire Prevention Bureau under the direction of the local fire marshal. For example:

> In one state, a fire marshal is responsible for enforcing all state laws regarding arson, and for providing state regulations and requirements for flammable liquid use and storage, as well as for schools, hospitals, hotels, etc. Two large municipalities within the state have local fire marshals who are in charge of the municipal Fire Prevention Bureau. A fire protection project in one of these municipalities under the jurisdiction of the local Fire Marshal (and therefore under the jurisdiction of the Fire Prevention Bureau) must obtain approval of plans from this agency. The Fire Prevention Bureau may even take precedence over

the local Building Code plan approval, especially if the project falls directly under the guidelines of the Fire Code requirements.

When it comes to fire protection and/or life safety, the authority and requirements of the local fire marshal acting as head of the Fire Prevention Bureau, backed up by the enforcement of law-even outside the scope and letter of the Code-take precedence when it comes to plan or installation approval.

The architect, engineer, and contractor are advised to become familiar with the state, county, fire district, and local authorities having jurisdiction within the parameters of the specific project, the state, and the local area of the project within the state.

Keep in mind that the fire inspector from the Fire Marshal's office, local or state, or the Fire Prevention Bureau, can cite installation deficiencies and violations that do not appear in the Fire Code, although these are usually of a minor nature. If the installation has not deviated from the plans approved by the Building Code Department and the Fire Prevention Bureau, these inspection citations can usually be corrected inexpensively and quickly.

Other Authorities Having Jurisdiction

Federal authority having jurisdiction, and the requirements of federal agencies, must be considered in a different environment, as the federal agencies and organizations have, for the most part, their own set of requirements and regulations. Copies of these standards, including the military standards are available at the Whole Building Design Guide Site (https://www.wbdg.org/).

The General Services Administration, the Department of the Interior, and the Department of Transportation also publish standards that cover their specific area of jurisdiction and authority. The Environmental Protection Agency (EPA) also publishes standards, and the EPA can greatly impact fire protection projects.

Water Department

It may seem strange to include the water department under authorities having jurisdiction, but this is an agency that must be given consideration. Two very important items fall under the jurisdiction of the water department, and both can have a definite cost impact on a project-a major cost impact if the project has already been installed.

One is metering. Some water departments require that fire suppression systems that take supply from a city water main have a dedicated meter on the supply line. Two major questions must be answered. First, does the water department in the area where the project is located require the metering of fire suppression systems? If the answer is yes, the second question is: do they only require the metering of small flows, and not the metering of fire flows when the system is operating?

If the requirement is only metering of small flows, the specifications will require a UL-listed detector check meter. Small flows are metered through a bypass with flow through the check restricted by a weighted check valve.

In some cases the water department will require metering of both small flows and fire flows. The specification must now include a full flow fire meter. These are large devices, UL-listed, with an OS&Y valve on either side of the meter so it can be removed for repair or replacement.

With the detector check valve, the meter is in a small line bypass around the check valve, and with two shut-off valves on either side of the meter, the meter can be removed for replacement or repair without shutting down the fire line supply to the suppression system.

With the full flow fire meter, the two control valves are installed so that the entire meter can be removed for replacement or repair. When this occurs, the fire line supply to the suppression system is shut down. To avoid losing supply to the suppression system, there is usually a bypass around the meter with a normally dosed control valve. If the full flow meter were removed, the bypass could supply water to the suppression systems to keep the fire protection in service.

Metering, if required, is an important specification item, and a quite expensive equipment and installation item, especially if it involves the full flow fire meter. It is well worth the time and effort to establish definite design requirements with the water department.

The second area where the water department is the authority having jurisdiction is backflow prevention. Most U.S. cities are now requiring backflow prevention on fire suppression systems that take supply from a potable water supply.

Similar to the metering requirements of the water department, each water department that requires backflow preventer devices has specific requirements for a double check backflow preventer or a reduced pressure backflow preventer.

Always determine if backflow prevention is required in the water district or water department where the project is located. It is then extremely important to verify which type of backflow preventer is required. The following is an actual example of the problems that can occur when all of the information is not obtained.

It was learned that the water district where the project was located required backflow prevention on automatic sprinkler systems taking supply from a city water main, but no one had obtained all of the water district requirements.

The specification writer specified a double check backflow preventer, and the designer of the preliminary plans for bidding designed around a double check backflow preventer. The successful bidder included the cost of a double check backflow preventer in his estimate and proceeded to install his device. After the installation was complete and ready for inspection, the code department, which was fully aware of the water district requirements, denied the occupancy permit. It was then discovered that the water district requirements allowed a double check backflow preventer on standard sprinkler systems, but the requirements mandated a reduced pressure backflow preventer when the sprinkler system was supplied through a fire pump, and this particular project sprinkler system had a fire pump. The cost impact, and the legal problems concerning who was to pay for the change, were overshadowed by the delay the client suffered because he was denied an occupancy permit for several weeks.

Summary

The purpose of this chapter has been to impress upon the architect, engineer, specification writer, and estimator the critical importance of defining who is really the authority having jurisdiction over the project, and not leaving a specification with generic wording encompassing a multitude of authorities so no one will be missed. Defining exactly what agency, individual, organization, or insurance carrier is the authority having jurisdiction on a specific project may require a time-consuming evaluation, but in the final analysis, it will prove to be worth every effort and every hour.

Many projects have suffered a severe cost impact, time lost, and schedule destroyed impact, not to mention damaged public relations as a result of a quick and dirty specification, where every conceivable authority having jurisdiction is listed without having determined if some on the list could come back to haunt the project, or someone was omitted from the list who will appear later on to devastate the installation approval.

Fire Suppression System Specifications

14

Construction Specification Standards

Fire protection specifications are extremely important documents-from the standpoint of both project performance and quality assurance. The fire protection specification tells the contractor exactly what the designer wants in the way of equipment, material, installation, and overall project quality.

According to the American Institute of Architects (AIA) Document A201–2007, the Contract Documents for a construction project consist of "the Agreement, Conditions of the Contract, Drawing, Specifications, Addenda…", as well as other miscellaneous documents associated with the contract between the project Owner and the Contractor hired to complete the work. Construction specifications become a part of the legal documents of the agreement and form a cornerstone of the project design. In fact, in most cases, the construction specifications override the project drawings in the event of conflicting information.

The purpose of construction specifications is to delineate the requirements regarding the materials, products, installation procedures and quality aspects involved with execution of the work and fulfillment of the contract. Specifications can be divided into three primary categories: performance, prescriptive and proprietary, which are described below (Noll).

Fire protection contractors may sometimes question why a certain product was specified, when an alternative would reduce costs. Fire protection contractors may also criticize specifications with the attitude that the designer is not so knowledgeable about specifying a fire protection system. In fact, many fire protection contractors do not realize or understand the engineer's role in preparing the specification and have the attitude that the specification was written at the last minute based on the first time the engineer had a chance to look at the project. These contractors do not realize the hours spent at meetings, conducting project research, coordinating the work with the other disciplines, and making calculations and evaluations-all of which went into the reasons for specifying exactly what is in the specification.

On the other side of the coin, most fire protection contractors are professionals in what they do, and are, of course, in business to make a profit. The engineer who writes the specification does not always realize that he or she prepares the specification in an entirely different environment from that of the contractor.

When the fire protection contractor receives the project specification, he must take this document written by a professional stranger with unknown fire protection expertise and convert it into a bid that will obtain the contract for his company, while complying with the specification, and with the codes and standards that he knows and works with on a day to day basis.

© Springer International Publishing AG, part of Springer Nature 2019
R. C. Till, J. W. Coon, *Fire Protection*, https://doi.org/10.1007/978-3-319-90844-1_14

This is not an easy task in a competitive marketplace: To come up with a low bid and still give the client a quality installation. It becomes very evident to the professional fire contractor when he reads the specification whether or not the engineer has done a thorough job of investigating all aspects of the project. If there are concerns, he may feel compelled to add in a cost allowance for extras. Engineers who prepare fire protection specifications would benefit if they spend a minimum of six months as a specification apprentice working with a fire protection contractor. This is one way to discover the cost impacts of potential plan revisions. This is not a criticism of the engineers who prepare fire protection specifications or the fire protection contractor. The point is that each side should understand the others concerns and objectives. The engineer who writes the specification should realize how the fire protection contractor uses it, and vice versa.

The design engineers specification is more than an exercise in demonstrating his knowledge of fire protection systems; It is a real-life document that should both provide the client with a quality project and give the installing contractor the opportunity to be successful. One thing for an engineer to remember: You cannot fool a professional fire protection contractor. These professionals know by reading the first paragraph of a specification whether or not the preparer knows fire protection systems and understands what fire protection contracting is all about. The fire protection specification writer, cannot con an experienced fire protection contractor with pretended expertise in the specification or with cover all statements. Designers should always keep in mind when preparing the specification that someone is going to read it who has a great deal more hands-on knowledge of fire protection systems, regardless of whether you have a P. E. after your name or other prestigious college degree hanging on the wall. This thought should keep fire protection specification writers honest, humble, and precise.

Specifications can be divided into three primary categories: performance, prescriptive and proprietary, which are described below:

1. Performance Specifications – Tells the contractor what the final installed product should be able to do.
2. Prescriptive Specifications – Detailed explanation of the materials the contractor must use, and the means of installing those materials
3. Proprietary Specifications – In this case a single approved product is specified for a particular installation. This may be due to the fact that the owner wants to maintain consistency of materials or may simply prefers a specific type of product. Also, in some highly complex installations where there is only one specific piece of equipment that will accomplish a specified task, a proprietary specification may be required, for example a specific water pump for use with a particular water mist system.

References

Under References, be sure to give the contractor the specific information he or she needs. Is the client insured by an entity other than Factory Mutual? If so, list UL and the National Fire Protection Association. The statement, authorities having jurisdiction, does not tell the contractor what specific requirements will apply. If the client is insured by Factory Mutual, so state in the specification, but don't add NFPA. The Factory Mutual Data Sheets that are equivalent to NFPA Fire Codes will provide the necessary requirements, and the FM Data Sheets do not always agree with NFPA Fire Codes in all areas. FM has some specific requirements and exceptions in their Data Sheets even when they use an NFPA Fire Code word for word.

Applicable Standards

Here again, many specifications will list the gamut of standards that are to be applied to the project. These usually include Underwriters Laboratories, Factory Mutual, and the National Fire Protection Association.

If this is a Factory Mutual insured client, list Factory Mutual Data Sheets and the Factory Mutual Approval Guide. The Factory Mutual Guide is equivalent to the publication that lists UL equipment, but it indicates only the products and systems that have the Factory Mutual approval. There are some products, systems, equipment, and devices that are FM Approved, but not UL listed. By indicating both of these approval bodies, the specification is allowing the contractor to use any item that is FM approved or UL listed on a Factory Mutual project. It is true that Factory Mutual will accept some items that FM does not include in their Approval Guide but are UL listed, but when this discrepancy becomes apparent on a project, it should be handled item-by-item with approval from the designing engineer and Factory Mutual. Make this list of Applicable Standards specific to the project.

Design Criteria

If the engineer has prepared a preliminary layout and sized the piping on this preliminary layout by hydraulic calculations, the specification should not lock the contractor into using this layout or the pipe sizing per se.

Sometimes it is necessary for the engineer to prepare a preliminary layout and to size the piping by hydraulic calculations so that the estimator can have a base to work from, or so that the structural engineers have an idea of piping weight. Nevertheless, the contractor should be given the option to redesign and recalculate the system subject to the approval of the engineer and in strict compliance with the specifications.

This flexibility may result in a more efficient and, for the client, less expensive system, because every contractor will be making an effort to comply with the specification, while providing a system that will make the bid competitive. As quoted previously, the old saying applies as well: "Give a plan to five sprinkler contractors, and you will get five different approved layouts."

Should all control valves be supervised? If so, make this statement, but follow it by specifying how the control valves are to be supervised.

NFPA provides several means for the supervision of control valves, and Factory Mutual usually has very specific requirements. Some specifications include the statement that the requirements of the public health authority having jurisdiction shall be determined and complied with regarding public water contamination. The fact is that the engineer must determine what is required or not required. When it is determined that back-flow prevention is required, state in the specification which type of backflow preventer is required. This should not be left up to the contractor, especially during the bidding period, and it is not a requirement to be determined by the estimator.

Some specifications use the statement: Cathodic protection is required. The engineer is the only qualified person to determine this requirement. If it is required, so specify; if not, so specify. The responsibility is in the hands of the specification writer.

Metering is another item that is the engineer's responsibility to determine. If metering is not required, so specify, but if it is required (based on investigation) specify which type of metering is required. Some specifications will include a section covering design of the hose stations. If the engineer has determined that hose stations are required, he should specify exactly what equipment shall be used, and how the hose stations are to be supplied. Are they to be small hose supplied from wet pipe sprinkler systems, or hose stations supplied from a standpipe that is also supplying sprinkler systems? Hose type and quality, type of nozzle, rack or reel, and hanging arrangements should all be stated. There are times when hose stations are not required. If this is the case, do not plug in a section in the specification covering hose stations if required state that no hose stations are required.

When the system is a dry pipe system, do not specify a quick opening device if required. The engineer should know approximately the volume of the dry pipe system, and whether or not the exhauster or quick opening device is required. The engineer has to know the approximate volume of the dry system in order to determine if one system will accommodate the project.

Always specify that the shop drawings, or installation plans, shall indicate the dry pipe system capacity. This is very important when the shop drawings are submitted for approval.

Hydraulic Calculation Specifications

As previously mentioned, design criteria must be determined by the engineer and clearly stated in the specification. If several areas of the project facility will have different design criteria requirements, they can be indicated on the plans, and the specification can read: "Design criteria relative to water density and area of operation shall be as indicated on the plans."

An important point, often left out of a specification for a dry pipe system, is the 30% increase in area of operation that may be required when hydraulically calculating a dry pipe system. If the specification reads simply so many gpm per square foot over so many square feet of operation, the contractor is not sure if the 30% is included or if he is to add it. This will make a big difference on his estimate sheet.

Velocity pressure is sometimes excluded by the specification, stating that velocity pressure shall not be used in the hydraulic calculations. Whether or not to allow the contractor to use velocity pressure in his calculations is sometimes a judgment call by the engineer, and sometimes a requirement of the insurance interest having jurisdiction. It is a good idea to verify the insurance carrier's requirements on this issue before it is written into or left out of the specification.

If velocity pressure is strictly a judgment call by the engineer, that judgment will have to be based on the project conditions, and whether or not this safety factor can be eliminated in the calculations. Water supply is a big factor and some water supplies, coupled with the specified system, will not be capable of satisfying the design requirements without using velocity pressure.

This discussion on velocity pressure also applies to velocity of the water in the piping. The velocity (feet per second) of water in the piping can also be left to the judgment of the engineer or can be a specific requirement of the insurance

carrier on the project. If the specification is going to restrict the calculations to a specific velocity, be absolutely sure that the velocity restriction so specified will meet the approval of the insurance carrier, and that the specified velocity has not defeated the calculation results relative to supply vs. demand.

Do not specify that velocity shall be in accordance with NFPA requirements, because NFPA does not provide requirements relative to velocity. Nor shall the specification writer state that velocity will comply with the insurance interest having jurisdiction. The engineer/spec writer must make this professional determination and so specify.

It is always good practice to include in the specification some wording that the installing contractor shall not add heads or institute pipe changes in direction or size without submitting revised calculations. This is a statement that will prevent the installing crew from changing piping layout in the field to avoid an obstruction, without realizing that these changes can affect the hydraulic calculations.

Acceptable Manufacturers

Specifications usually include a statement referring to all those manufacturers listed in the UL publications, and all those listed in the Factory Mutual Approval Guide. Some manufacturers products that are listed in the UL publication are not listed in the FM Approval Guide. One example is a commonly used valve in fire protection systems. It is UL listed, but is not FM approved. If the project is FM insured, this specification gives the contractor the option to use either valve, which may lead to a denial of FM approval of the system. If the project is FM insured, use the Factory Mutual Approval Guide. If the project is not an FM risk, use the UL publications. As mentioned before, on an FM project the contractor can usually get FM to approve the use of a UL listed product if for some reason the FM approved product is not available, but the contractor should not be given an open choice on an FM project.

Equipment and Material

Here again, specifications usually include the statement that all equipment, devices, and accessories shall be either UL listed or FM approved. Determine which is applicable and so specify. It is good practice to limit the contractor to the use of products from a single manufacturer when two or more products of the same type are required by the system or systems. Do not allow the contractor to mix and match with products he might have on the shelf. This does not produce a quality installation, and when replacement parts or units are required, it can cause additional problems.

Most specifications will leave the galvanizing or corrosion protection of piping up to the contractor by stating that the piping shall be protected in areas that require it. It would be a much better and more professional practice for the engineer to determine where these areas exist, and then either indicate them on the plan or identify them in the specification. This is really the responsibility of the engineer who should be quite familiar with the entire project and project occupancy. The desired trim and accessories should be specified for valves, alarm valves, dry pipe valves, and deluge valves. This applies not only to the standard trim that comes with the complete valve, but anything that might be considered extra to the standard trim. For example, a retard chamber, a pressure switch, a pressure switch with two sets of contacts, a water motor alarm, an electric alarm, etc. This is good practice, because it requires the estimator to include all these items in his estimate. If a dry pipe valve house is required, it should not be included in the fire protection contractors specification unless it is intended that he include the construction in his bid. Usually this house will be designed by an architect and should be specified in the general construction specification. This is mentioned because some specifications will state that the dry pipe valve shall be installed in a dry pipe valve house, and the fire protection contractor is not sure whether he should include this in his bid, or at least the equipment that must be installed in the valve house, such as the thermostatically controlled heater and low temperature alarm.

If the system design calls for a vane-type water flow switch, be absolutely sure to specify that the flow switch shall be the instantly recycling type, so that flows that do not activate the device (because they occur within the retard time frame) are not cumulative. The water flow switch should have a field adjustable retard, but it is the responsibility of the engineer preparing the specification to determine to the best of his ability the retard time. Make sure by specification and/or plan detail that the required pressure gauge and main drain are included in the water flow detector riser.

Sprinkler Heads

The usual statement in specifications that all sprinklers shall be UL listed or FM approved, a blanket statement, is not so problematic when referencing sprinkler heads used on standard sprinkler systems, but problems do arise when the system involves a newly developed head that may have a UL listing, but has not yet been approved by Factory Mutual. Also, in some systems, foam systems for example, sprinkler heads have FM approval, but no UL listing. To be more professional, and subject to fewer conflicts, the engineer writing the specification should be sure the heads on his or her project carry both the UL listing and FM approval.

If there are suspended ceilings with concealed piping and pendent heads, and the heads are to be a special type, it may be necessary to specify a manufacturers model number or engineer approved equal. This is because a pendent head specified as a flush head does not necessarily mean that the specifier will get the type of flush head expected. There are several types and patterns of flush heads, and several that are referred to as flush heads that are really not what one might expect.

The decision to require corrosion-proof heads should be the responsibility of the engineer. He or she has lived with the project since its conception and should know where these corrosive problem areas exist. The contractor should then be informed by specification and/or plan notations. When the shop drawings are received for review, or when the installation is inspected for final approval and certain areas are recommended to have corrosion-proof heads, a specification item will eliminate any discussion. If the as required was the specification item, the engineer could be faced with a difference of opinion with the contractor.

Make sure the specification for spare heads includes the requirement that spare heads of every type used on the project be included in the spare head cabinet, and that a sprinkler wrench specifically adapted to removal and replacement of every type of head used on the project be included in the spare head cabinet.

Temperature rating of sprinkler heads can be a tough call in a specification. The engineer should know which areas will require high temperature heads and should specify these areas or indicate them on the plans. One situation that does not require a judgment call is when the insurance carrier mandates the temperature rating of the heads; this should definitely be specified. In new buildings, the engineer should have a fairly good idea of where unusual heat conditions will exist, or where heat producing equipment will be located. This information should be included in the specification or so noted on the plan. Unless the insurance carrier requires a specific temperature rating, NFPA and FM offer excellent guidelines relative to head versus ambient temperature conditions.

If there is no way to determine the maximum temperature of an area, it is a good idea to have the contractor furnish and install a recording thermometer. The resulting temperature readings will determine the proper head temperature. This may require a return trip to the job by the contractor, with ladders and a crew if the installation is complete before the temperature determination is made. This extra expense should be discussed and agreed upon by all parties concerned, since it could have been avoided if the engineer had determined the heat conditions and head temperature rating when the specification was prepared during design.

Be sure that the specification for the type of heads does not conflict with NFPA, FM, or the insurance interest having jurisdiction.

Drains and test connections are not a major specification problem, but if the specification states that the drain shall discharge to the closest drain outlet, and the engineer knows that the closest drain outlet is a floor drain, be sure to specify that the drain line be carried to within, say 6" of the floor drain. Otherwise, some drain discharge lines will spill out on the floor far from the floor drain, and result in water damage, or at least an unnecessary mess. The sprinkler engineer should always inform the plumbing engineer of this drain discharge and the amount of water anticipated, so the floor drain can accept the volume without causing an overflow.

Drains tend to be bypassed on plans and in specifications. There is a sort of a leave-it-up-to-the-contractor attitude. This is definitely wrong because unless the contractor is extremely conscientious, he will be inclined to arrange the drain discharge in the least expensive manner. Specify exactly where the drain should be located, and where it should discharge.

Under the category of control valves, there are one or two cautions. When specifying wall post indicator valves, be sure to specify the specific type. There are recessed, fully recessed, and totally exposed wall post indicators. Depending on which type is desired, the engineer must coordinate the design and specification provisions with the architect, and possibly the structural engineer, depending on the type of wall structure involved.

When a wall post indicator is to be used on a project, always specify, or better yet detail on the plan, the location of the flange and spigot piece off the wall. Each type of wall post indicator requires a certain distance off the wall to facilitate installation.

And speaking of the flange and spigot piece, specify that the flange shall be located 6"–8" above the finished floor. Having the flange of the flange-and-spigot piece a foot or so above the finished floor is not acceptable.

All control valves on the site of a facility should open in the same direction. If the installation is an addition to a sprinklered facility, determine which way the existing valves open, and specify that the new valves open in the same direction. This is extremely important for outside control valves installed on the underground. If a valve is turned in the wrong direction to open, it can be seriously damaged, and may become inoperable.

Pipe supports and attachments must be UL listed or FM approved. Approved standard hanger requirements are adequately covered by NFPA and FM publications. There must not be any conflict in the specification between a listed or approved fire protection pipe support and a hanger designed by an engineer who has used his hanger arrangement on process piping. The hangers for fire protection must be UL listed or FM approved, including rod size, or the system will not comply with NFPA or FM requirements.

The engineer should recommend in the specification that no sprinkler piping be supported from the bottom chord of bar joists, because too much weight on the bottom chord of the bar joist can turn the bar joist out of the vertical position. Always include an earthquake zone hanger requirement, and when a fire department connection is specified, always include the statement that the hose threads shall comply with the specifications of the local fire department. Or obtain from the local fire department their hose thread specifications and include those in the specification. Know your project thoroughly before you specify the fire department connection, and then specify exactly what is needed on the specific project. Always specify that the ball drip drain valve be installed to maintain the fire department connection in a dry state. Be aware that this automatic ball drip drain will, at times, discharge a small amount of water and provide for this occurrence.

Testing

Specify that all tests shall be conducted by the installing contractor and specify who shall provide all labor and equipment to conduct these tests. Always specify that the contractor shall notify all concerned parties (owner, engineer, insurance carrier, fire marshal) five days or more in advance that the tests will be conducted. If the situation requires more advance notice, inform all concerned parties accordingly so that they

may witness the test. This notice should also be issued several days before the flushing of the underground, so that this operation can also be witnessed by all concerned parties.

Specify that following the testing of a system, the system shall be returned to a functional and operational condition at no extra cost to the owner. Any retesting that is required because of failure of any test shall be specified to be conducted at no additional cost to the owner, and any corrections or repairs to the system to permit retesting shall also be performed at no cost to the owner. Unless the specification is complete, specific, and tight, the engineer will not have a leg to stand on if there is a dispute with the contractor and/or client when the shop or installation plans are reviewed for engineer approval, or when the installation is complete. Field supervisory personnel need a good, well prepared, specification in order to make sure the contractor is complying with the specifications. A specification that merely states that the system shall be installed in accordance with NFPA could be in for a lot of problems and cost impacts. The problem is not that NFPA is not an effective guide, but rather that there are many areas where a contractor can take advantage of a less expensive arrangement that meets NFPA requirements, but definitely does not meet the client's desires.

Summary

The fire protection specifier must be made aware of the serious problems that can be caused by blanket statements in a specification. This individual must not only thoroughly understand, but also convey, an understanding of the project requirements and authorities having jurisdiction, and of exactly what is required on the project.

Reference

Noll, Michael. "Types of Construction Specifications." 3/13/18 <https://www.archtoolbox.com/representation/specifications/types-of-construction-specifications.html>.

Printed in the United States
By Bookmasters